# MANAGING
# NEW INDUSTRY
# CREATION

THOMAS P. MURTHA,

STEFANIE ANN LENWAY,

and JEFFREY A. HART

Finally, a study that offers critical new insights into the complex issue of the evolution of global high-tech industries. This scholarly and painstakingly researched study of the flat panel display industry breaks new ground and represents a clear departure from traditional economic analysis. The authors have developed a fundamentally new and managerially relevant perspective by combining analysis of knowledge creation and industry evolution—people, cross-national teams, investments, and public policy—and by demystifying the glue that binds the apparently distinct roles played by firms and countries; a required reading for those who are concerned about innovation and competition in global industries. —**C.K. Prahalad**, chairman, PRAJA, Inc., Harvey C. Fruehauf Professor of Business Administration, University of Michigan Business School, and coauthor, *Competing for the Future.*

Five years of intensive research and conscientious interaction with the display community have produced the only detailed, readable history of the flat panel display industry from early research in the 1960s to the turbulent, high-volume present. Some of the authors' conclusions will be controversial, but noncontroversial conclusions are a waste of time. This book is certainly not a waste of time—it is essential reading for everyone in the display industry and for anyone with a stake in global business management. —**Ken Werner**, president, Nutmeg Consultants and editor, *Information Display* magazine.

This book is a fascinating analysis of the emergence of the first "born global" industry, flat panel displays. It highlights the limits of both conventional business strategy thinking and traditional public policy making in helping us understand the emergence and development of global industries, and points to the dynamics of new global knowledge alliances that underpin the creation of new industries. —**Yves L. Doz**, Timken Professor of Global Technology and Innovation, Dean of Executive Education, INSEAD, and coauthor, *Alliance Advantage* and *From Global to Metanational: How Companies Win in the Knowledge Economy.*

In *New Industry Creation*, Murtha, Lenway, and Hart closely examine the birth process of a single high-tech industry. In this fascinating account, they draw valuable lessons about knowledge transfer across geographical boundaries, successful and unsuccessful interplay between research, product development, and manufacturing, and the role of management and governments. There is much here that applies to other industries, and we are reminded again that while sound strategies are important, in the end, details of execution matter. —**Frank Mayadas**, program director, the Alfred P. Sloan Foundation.

This book is the definitive work on the emergence of the flat panel display industry. During the critical growth years, the authors conducted numerous in-depth interviews with the major players around the world. The result is a comprehensive picture of the real factors that determined the eventual winners, as well as valuable insights that apply to today's fast-moving, knowledge-based industries. —**Steven W. Depp**, director, subsystem technologies and applications laboratory, IBM (retired).

This book is a rich and valuable contribution. It explodes many old myths and stereotypes about technology development in an increasingly internationalized business environment. This is must reading for academics, policymakers, and thoughtful business executives in the United States and around the world interested in technology development and management, organizational behavior, international learning, and knowledge creation. —**Martin Kenney**, professor, human and community development, University of California, Davis, editor, *Understanding Silicon Valley*, and coauthor, *The Breakthrough Illusion.*

# MANAGING NEW INDUSTRY CREATION

GLOBAL KNOWLEDGE FORMATION
AND ENTREPRENEURSHIP IN
HIGH TECHNOLOGY

THOMAS P. MURTHA,

STEFANIE ANN LENWAY,

and JEFFREY A. HART

STANFORD UNIVERSITY PRESS
*Stanford, California*

Stanford University Press
Stanford, California
©2001 by the Board of Trustees of the
Leland Stanford Junior University

Printed in the United States of America on acid-free,
archival-quality paper.

Library of Congress Cataloging-in-Publication Data
Murtha Thomas P.
    Managing new industry creation: global knowledge formation and entrepreneurship in
high technology: the race to commercialize flat panel displays/Thomas P. Murtha,
Stefanie Ann Lenway and Jeffery A. Hart.
*JK*     p. cm.   (Stanford business books)
    Includes bibliographical references and index.
    ISBN 0-8047-4228-6 (alk. paper)   ISBN 0-8047-4246-4 (pbk : alk. paper)
      1. New products—Management. 2. New Products—Marketing. 3. High technology
industries—Management. 4. Technology transfer. 5. Globalization—Economic aspects. I

    HF5415.153 M87 2001
    658.5'75—dc21                                              2001049225

Original Printing 2001

Last figure below indicates year of this printing:

10   09   08   07   06   05   04   03   02   01


Typeset by Interactive Composition Corporation

# CONTENTS

## APPENDIXES

# TABLES AND FIGURES

In Memory of Raymond Vernon
1913–1999

# THE AUTHORS

**THOMAS P. MURTHA** is associate professor of strategic management and organization at the University of Minnesota and managing partner in Stratametrics, Inc., a global management consulting firm. He received his Ph.D. from New York University and has served on the faculty and as coordinator of doctoral studies in international business at the University of Michigan Business School. Previously he worked in the New York advertising industry, serving as an executive vice president in a number of agencies with mainly Japanese consumer electronics firms as clients. In addition to academic articles on international business strategy in such publications as *Strategic Management Journal* and *The Journal of Law, Economics and Organization,* he has written on the arts and consumer electronics for leading newspapers and magazines, including *Rolling Stone* and the *Minneapolis Star Tribune.* He is an associate editor of the *Journal of International Business Studies.* Tom Murtha can be reached at tmurtha@csom.umn.edu.

**STEFANIE ANN LENWAY** is professor and department chair of strategic management and organization at the University of Minnesota, and president of Stratametrics, Inc. She received her Ph.D. from the University of California, Berkeley. Lenway has served as chair of the Academy of Management's Social Issues in Management Division and as vice president of the Academy of International Business; she is a member of the Board of Governors of the Academy of Management and an associate editor of the *Journal of International Business Studies.* Lenway's previous book was *The Politics of U.S. International Trade Policy.* She has also written many academic articles on strategic management, politics, and economics for such journals as *Strategic Management Journal, Academy of Management Journal,* and *International Organization.* Stefanie Lenway can be reached at slenway@csom.umn.edu.

**JEFFREY HART** is professor and department chair of political science at Indiana University, Bloomington, where he has taught international politics and international political economy since 1981. Hart received his Ph.D. from the University of California, Berkeley. He taught at Princeton University from 1973 to 1980 and was a professional staff member of the President's Commission for a National Agenda for the Eighties. Hart has worked at the Office of Technology Assessment of the United States Congress, helping to write *International Competition in Services* in 1987. He has been a visiting scholar at the Berkeley Roundtable on the International Economy. Hart's authored, coauthored or co-edited books include: *The New International Economic Order, Interdependence in the Post Multilateral Era, Rival Capitalists, The Politics of International Economic Relations,* and *Globalization and Governance* as well as articles for *World Politics, International Organization, The British Journal of Political Science, New Political Economy,* and *The Journal of Conflict Resolution.* Jeffrey Hart can be reached at hartj@indiana.edu.

# PREFACE AND ACKNOWLEDGMENTS

*To accept some measure of confusion about
one's national identity is not an
easy thing...*

RAYMOND VERNON, *SOVEREIGNTY AT BAY,*
1971, p. 109

The growing knowledge-intensiveness of global economic activity demands new ways of thinking about industry, competition, and strategic management. This need presented itself to us dramatically in the research for this book. Our research started out as an investigation of an emerging high-technology industry that, for many observers including ourselves, represented a crisis of competitiveness for U.S. companies. The genesis of the book occurred in our discovery that we were wrong. We had focused on the accumulation of physical plant and equipment, at the time concentrated in Japan, as the essential dynamic that defined new industry creation and its management challenges. In fact, the essential dynamics to be managed were global learning and knowledge-creation processes that necessarily engaged an international community of companies. Along with Japanese competitors, alliance partners, suppliers, and customers, U.S. companies with strong organizational capabilities in Japan played essential roles.

This book distills principles for strategy in knowledge-driven competition from managers' experiences in the companies that created the high-volume flat panel display (FPD) industry. Our objective has been to derive something quite general from the rich, multicultural specifics of this industry's emergence, which we have been fortunate to observe first-hand for almost the entire decade since large-format color FPDs were first commercialized. As a reader, we ask that you set aside any preconceptions you may have that a book about a single industry can address only the narrow concerns of specialists. We can offer at least four reasons why the book will interest you as a manager, R&D scientist, consultant, public policy-maker, academic, student, general reader, or all of the above if you are concerned about technology, innovation, and global economic change.

First, although the flat panel display industry's history and individual personalities in themselves make a compelling story of vision, risk-taking, and invention, we have presented more here than history. We have tried to show how unique aspects of the FPD industry's emergence foreshadowed challenges that will define new industry creation in the future, most notably globalization from the start combined with an unprecedented rate of change. Rapid change was both a cause and a consequence of knowledge-intensity. The central strategic challenge involved managing people and partnerships internationally, over time to repeatedly create, retain, and transfer knowledge across new product and manufacturing generations before anyone else could do it.

Second, the book contains policy rationales for both business and government decision-makers concerned with maintaining a healthy technology base for countries in a world where innovation increasingly depends on access to global knowledge-creation processes.

Third, the book will hold intrinsic interest for readers interested in advanced person-machine interface research as well as long-standing R&D efforts to develop large, wall-hanging flat TVs. We have examined the creative processes that underlie both the technology and the organizational innovations driving these efforts forward.

Finally, we are unaware of any book or other report that collates the experiences of successful Japanese, U.S., Korean, European, and Taiwanese flat panel display industry participants, almost all of whom we visited during our fieldwork. We learned that the industry's early concentration in Japan intensified in part because individuals, teams, and organizations needed to interact at first hand to keep pace and contribute meaningfully to the high rate of knowledge accumulation. The same held true for our efforts to understand what was happening.

We have organized the book in nine chapters. Chapter 1 provides a conceptual overview of frameworks for strategic and organizational analysis that we use throughout the book to explain industry evolution. We introduce *continuity, learning,* and *speed* as core dimensions of successful strategies in knowledge-driven competition. Chapter 2 provides a narrative overview that will orient the reader in the historical chapters that follow.

In Chapters 3 through 6, the industry's history unfolds more or less chronologically. Chapter 3 pertains to the period from the invention of basic technologies to the announcement of the first large-format color flat panel prototypes in Japan in 1988. We show how companies that implemented management and technology processes based on *continuity* and *learning* qualified themselves for leadership. Chapter 4 describes the interregnum between the prototype announcements and companies' decisions to make the costly investments in fabrication lines and manufacturing processes necessary to commercialize the large displays. In the companies that proceeded, senior managers set aside negative recommendations based on financial models because they envisioned displays as keys to future markets that could not be quantified. Chapters 5 and 6 examine the torrid interaction of product and process innovation in Japan during the 1990s, where companies vied to enlarge the display sizes they could manufacture, while at the same time driving costs down to reach a mass market. *Learning* and *speed* dominated companies' efforts to confront the paradox of increasing investment requirements combined with the need to decrease prices.

Chapters 3 and 5 also include boxes set off from the main text that explain alternative FPD technologies and manufacturing processes in plain language.

In Chapter 7, we build on the narrative of previous chapters to discuss the frameworks touched on in Chapter 1 in richer detail. Chapter 8 examines issues that affected the industry's diffusion to manufacturing locations outside of Japan, particularly the relative openness of corporate and government policies and practices toward global knowledge-creation partnerships. Chapter 9 offers a conclusion that extends our perspective on knowledge and new industry creation to other industries and brings the flat panel display industry narrative up to date.

## ACKNOWLEDGMENTS

In introducing a book such as this, authors customarily recognize debts, intellectual and otherwise, incurred in the process of getting the work done. The idea of debt implies, however, that an obligation has been incurred that can somehow be repaid. This is rarely possible, and it is certainly not possible in the case of C.K. Prahalad's contributions to our thinking, which he freely offered in many early discussions as we framed our research plans. C.K.'s comments on drafts of research protocols, early chapters, and our original book proposal echo through the completed work.

Yves Doz brought the same acuity to discussions in the middle phases of our research and read the manuscript as it neared completion. Yves challenged us to deeper thinking and higher levels of accomplishment. We cannot overstate our appreciation.

We are grateful to the Alfred P. Sloan Foundation's Industry Studies Program for the funding that made this research possible. Our program director, Hirsh Cohen, motivated us with his curiosity about the outcome of our research and built our confidence with his warmth, humor, and high expectations. Perhaps most important, Hirsh encouraged us to take risks. Our first proposal outline centered on a statistical model of industry evolution, to which he responded, "Don't you people have tenure?" As Hirsh moved toward retirement, Frank Mayadas sustained our relationship with the foundation and kept up the encouragement.

Our editor at Stanford University Press, Bill Hicks, was the final catalyst in the process that produced this book. We have benefited from Bill's long experience as an editor specializing in the management field, but probably appreciate his literary sensibility even more. It is good to write for an editor who loves writing and writers.

Many others played important roles in helping to get the research off the ground. Michael Borrus offered the facilities, Silicon Valley credibility, and intellectual firepower of the Berkeley Roundtable on the International Economy (BRIE). David Mentley offered early contacts and technical briefings. Michael F. Ciesinski, along with the members of the United States Display Consortium, provided open, generous interview access, as well as speaking opportunities in which we could test some of our early ideas. Andrew H. Van de Ven sparked our interest in innovation as a research field and made a vital connection by introducing Stefanie to Hirsh Cohen.

As our research progressed, we benefited from the comments of numerous colleagues and seminar groups. Martin Kenney read the manuscript and offered unrelenting encouragement. With Richard Florida, Martin coordinated The

Alfred P. Sloan Foundation globalization study group, which heard numerous reports of our findings. As director of the Strategic Management Research Center at Minnesota, Bala Chakravarthy sponsored, read, and cheered us on. He also exposed our work to astute criticism in several senior management workshops. Tom Levenson took time away from his own book, *Einstein in Berlin*, when it was already behind schedule, to comment on the early chapters of ours.

Tadao Kagono invited us to give a workshop at Kobe University and inspired us with comments that cut to the heart of knowledge creation as a perspective on new industry creation. He also helped managers at a particularly central, but "shy" company to overcome their reticence to meet with us. Without Kagono, there would be a tremendous void at the heart of our story.

In Korea, Chung In Moon invited us for a workshop at Yonsei University and provided us with background and introductions. In Tokyo, Johan Bergquist and the Asian Technology Information Program (ATIP) provided background, introductions, and the dais for two press and analysts' briefings, the first of which unleashed an uproar among several participants that two of us will never forget.

Many more friends and colleagues stood ready to break down barriers and raise up challenges at times when we really needed someone to do one or both of these things. Among them we include Kaz Asakawa, Stephen Boyd, Larry Davidson, Peter Dwyer, Rob Ehle, Judith Hibbard, Mark Freeman, Glenn Fong, Paul Johnson, Mike Houston, David Kidwell, Robert T. Kudrle, Sun-Kyoo Lim, Wallace Lopez, Alfred Marcus, Vladimir Pucik, Peter Smith Ring, Paul Semenza, Myles Shaver, Kate Wahl, Ken Werner, Eleanor Westney, A Williams, Aks Zaheer, Sri Zaheer, and Mahmood Zaidi. David Lamb and the Honeywell Technology Center funded a supplemental grant that helped seed the project. Ross Young and DisplaySearch provided no-strings-attached access to important data for profiling the industry.

We also benefited from seminars at INSEAD, Osaka University, New York University, Wharton, the University of Michigan, The International Trade Policy Workshop at Minnesota's Hubert H. Humphrey Institute, the European International Business Association, DisplayWorks, and the 1999 Economics of the Display Industry Conference.

Capable doctoral students assisted us at each of our institutions. In Minnesota, Jennifer Wynne Spencer's invaluable 1997 dissertation, *Firms' Strategies in the Global Innovation System*, grew out of our project and mounted a successful challenge to some of our most cherished early assumptions. Scott Johnson and Sharon James Wade batted cleanup in the closing days of the writing. Greg Linden assisted us from BRIE, coordinated our efforts, collaborated on interviews in Taiwan, and has since kept us on our toes with continued discussions of current industry developments. At Indiana, Sangbae Kim coordinated our Korean expedition, and Craig Ortsey also worked on the project.

We heartily thank Sharon Hansen of the Strategic Management Research Center at Minnesota for administering our Sloan Foundation funds for three of the project's four years, keeping us free of administrative complexities that would otherwise have distracted us from the work.

As we conclude these last few paragraphs in a project that has occupied many years, our thoughts turn to Raymond Vernon, who inspired us with his writing and personal example of scholarship, taught some of our most inspiring teachers and, indeed, many of our teachers' teachers. Raymond Vernon is, unbelievably, gone, but it is life-affirming to look back on his 1971 book, *Sovereignty at Bay*, which probably brought as many aspiring international managers and academics back to graduate school as Helen's face sailed ships to Troy. Having at last finished our book, we realize that much that we have written can be summarized in one of Ray's trenchant observations:

> Because so many of the problems of large corporations are those entailed in adaptation and change . . . emphases tend to appear that are not consistent with the usual profit-maximizing yield-on-investment model. Among other things, the management of the enterprise will generally regard its fixed assets as less unique and less difficult to reproduce than the people and practices and collective memory that comprise its organization . . . (p. 117).

In reading the above for the contemporary, knowledge-driven economy, as perhaps in 1971, we might infer a subtext that would replace "generally" with "ideally," or "enterprise" with "successful enterprise." Either way, it is no less true.

*April 2001*

THOMAS P. MURTHA
STEFANIE ANN LENWAY
*Gold Mountain, California*
JEFFREY A. HART
*Bloomington, Indiana*

# 1

# Industry Creation as Knowledge Creation

*Sharp Develops Thin, 14-Inch TV Monitor*
*Light Enough to Hang on Wall*
HEADLINE, ASAHI NEWS SERVICE,
JUNE 17, 1988

*Toshiba, IBM Claim Largest*
*Color Liquid Crystal Display*
HEADLINE, JAPAN ECONOMIC NEWSWIRE,
SEPTEMBER 21, 1988[1]

The information age has not ended the age of great new manufacturing industries. Since the early 1960s, scientists and product developers have worked to perfect display technologies that would make giant, wall-hanging, flat TV sets possible—and affordable—for any household. When 14-inch working prototypes of color liquid crystal displays (LCDs) were demonstrated in 1988, this dream crossed into the realm of possibility. Yet the new high-information-content flat panel displays (FPDs) did not at first capture public imagination in TV sets, but as a critical enabling technology for notebook computers, mobile Internet access, and pervasive digitalization.

If, as the saying goes, eyes are the windows to our souls, FPDs are the windows to the souls of our new machines.[2] The new machines will include countless products that will define the twenty-first century, including giant, wall-hanging, high-definition televisions, but also wearable computers, on-board automobile navigation, and remote medical imaging. FPDs will function as the face by which technology products will be known, convey meaning, and interact with users. By the early 1990s, every high technology CEO on earth knew that the future of his or her company in some way involved FPDs. FPDs had joined batteries, integrated circuits, and memory as the fourth essential hardware ingredient for competing on the digital edge.

The origins of this new industry marked a paradigm shift in the geography of innovation. Yet more than a decade later, the outlines of that shift and its implications for corporate strategy and government policy remain largely misunderstood. Old habits of thinking die hard. For many years, the most significant new high-technology industries, such as videotape, integrated circuits, and software, emerged first in the United States from ideas fostered in U.S. companies and research labs. When international competition broke out in these new product markets, the privilege of defending—and losing—market share belonged almost entirely to U.S. companies. In several high-profile instances, most notably dynamic random access memory chips (DRAMs), competition from Japanese and Korean companies nearly squeezed U.S. companies out. These experiences contributed to an atmosphere in which businesspeople, policy-makers, media, and academics often framed global high-tech competition as competition among countries and world regions rather than competition among firms.

## KNOWLEDGE-DRIVEN COMPETITION AND GLOBALIZATION

The FPD industry represents an example of a new class of industry—a class that includes software, e-commerce, and multimedia—that cannot be understood in terms of country rivalry or national competitive advantage. In these knowledge-driven industries, learning rather than product positioning powers competition. The knowledge-creation processes that fuel these industries' advances cannot be segregated by national territory. Consequently, companies can no longer project themselves from nation-centric strongholds that function as containers of competitive advantage. Instead, they need strategies that globally mobilize country-specific strengths and freely leverage them with partners, suppliers, and customers. Access to knowledge-creation processes matters more than ownership of physical assets. Physical location matters only insofar as it confers learning advantages and market access.

FPD technology commercialized first in Japan. The industry emerged from humble beginnings in the early 1970s when relatively small Japanese companies based new watch and calculator display applications on FPD technologies that largely originated in the United States. In 2000, nearly half of FPDs were still manufactured there and almost none in the U.S. or Europe. Although Japanese companies produced the first consumer products using FPDs in the 1970s, the manufacturing platform that made high-volume production viable in the early 1990s depended on joint efforts among companies from several countries. These companies got into the business because senior management envisioned the future importance of FPDs as a nodal technology in a global multimedia market of incalculable size. Otherwise, they could never have justified the immense investments required to achieve high-volume production of the most advanced displays. This convergence of factors—an international community of companies leveraging unique, national capabilities from several countries in a single locality to create a new global industry—had no precedent in the history of technology. The FPD industry was the first new manufacturing industry to fully emerge in a global economy defined by trade in knowledge rather than trade in physical products.

## GLOBALIZATION AND INTERDEPENDENT KNOWLEDGE CREATION

These unique aspects of the FPD industry were not evident in the early 1990s. Between 1991 and 1996, at least twenty-five high-volume fabrication lines were started up to produce the most advanced FPDs; twenty-one of them in Japan.[3] Judging strictly on the basis of physical plant locations and product origin, many U.S. business, government, media, and academic observers raised alarms about a threat to U.S. competitiveness. What would prevent Japanese companies from limiting the access of non-Japanese companies or even governments to the leading-edge FPDs they might need to stay in the technological forefront? Other observers focused on the torrid pace of technological evolution, declining prices, and cyclical over- and under-capacity to assert that the FPD industry was barreling headlong toward a mature, commodity market status where permanent oversupply would prevail. In the first scenario, non-Japanese companies would be forced to accept dependent relationships with Japanese production sources, unless plants were erected elsewhere in the world (preferably the United States). In the second scenario, non-Japanese companies could maintain independent relationships with numerous Japanese FPD suppliers desperately undercutting each others' prices to fill excess capacity for more-or-less identical products. The idea of building plants anywhere seemed quixotic.

Neither scenario envisioned the interdependent, global industry structure that was, in fact, emerging. Two related phenomena propelled this global interdependence. First, companies needed continual access to the rapidly evolving markets for products that incorporated FPDs. Second, they needed to participate in the rapid pace of knowledge accumulation and change in FPD technology itself. No single country held a monopoly of the resources that would enable companies to meet these needs. The elements needed to create the industry did not exist together in any one country, but were scattered across the globe.

The forces driving the first phenomenon—changing markets for the products that ignited FPD industry growth—were global and not centered in Japan, although Japanese consumers and companies were significant participants. The companies holding the preponderance of market share for products that incorporated the most advanced displays were not Japanese. Three of the global top-four notebook sellers (typically Toshiba, IBM, Dell, and Compaq) have consistently been U.S.-based. Once the notebook market emerged, it evolved rapidly due to the increasing size and sophistication of the mobile workforce, rapid progress in personal computing technology, the emergence of the World Wide Web, the spread of high-speed Internet access, and the rapid expansion of digital infrastructure. Most of the organizations whose systems and solutions impelled these changes—including Intel, Microsoft, and the Defense Advanced Research Projects Administration (DARPA)—were from the U.S. Microprocessors speeded up, the visually rich, icon-driven Windows operating system came to dominate desktops and the Internet morphed from an esoteric network linking scientists and academics into a mass culture phenomenon. Software achieved increasing levels of finesse.

Acting as the primary interface between all this computing power and the expanding market of users, FPDs required unremitting improvements in functionality. Producers improved their products by increasing screen size, clarity, sharpness, viewing angle, brightness, lightness, and video capability as well as decreasing power consumption. Product innovations required corresponding process innovations to drive costs down to expand the market and because each new technological advance created new manufacturing challenges. Pushing the technology into the future in the face of these evolving global requirements required learning partnerships that cut across company and national boundaries.

The forces driving the second phenomenon—the rapid pace of knowledge accumulation in FPD technology—were global, but ultimately caused production to center in Japan, with many non-Japanese participants. The high-volume industry emerged when it did through the interplay of internationally dispersed, knowledge-based competencies held by companies from several countries, most notably Japan and the United States. We will explain in this book how learning intensity, continuity in knowledge creation, and a torrid pace of change worked together to create a powerful, self-reinforcing, geographically focused cycle of technological progress.

The pioneering companies helped to create the preconditions for this cycle by implementing strategies based on criteria of *continuity* and *learning*. As the industry caught fire, they needed to race to stay ahead. The requirements for simultaneous product and process innovation evolved too rapidly for them to profit from traditional product-driven strategies of differentiation or cost leadership. Companies needed knowledge-driven strategies that simultaneously drove product innovation forward and costs down more quickly than could their rivals. *Speed* assumed equal importance with learning and continuity as the technology vaulted like a crowning forest fire from generation to generation. In the early days of industry evolution, only the fastest companies profited from change.

The interaction of increasing speed and learning intensified tendencies for the industry to concentrate in Japan in at least two ways. First, as the flow of discrete technological advances and other critical events in the industry's history compressed in time, first-hand observation played an increasingly vital role in assimilating new information. Yet the value of new information degraded ever more rapidly. At best, failures in timely observation created losses of continuity in the cumulative processes that underlie knowledge creation. At worst, these failures led into vicious cycles in which successive events progressively overwhelmed increasingly poorly informed companies or industry sectors. We encountered instances in which companies fell wildly out of phase with the state-of-the-art in FPD product and manufacturing technology. Some funded research projects to solve technical problems for which commercial solutions existed in the market. Others made arbitrary guesses about FPD dimensions that new generation manufacturing equipment and materials should accommodate, when in fact leading producers had already committed to specific plans.

Second, individuals working in the industry could spare less time to make their personal or tacit knowledge explicitly known to others, while the absolute quantity of personal knowledge employed in the industry was rising. The accumulation of explicit knowledge in the industry also increasingly outran the capabilities of companies to have it written down or otherwise codified for broader dissemination within their organizations or to customers or collaborators. Language differences amplified this gap for companies outside of Japan. Important documents in Japanese, for example, generally did not become available in English until at least one year had passed.

The difficulty of remaining informed and the shifting balance among tacit and explicit forms of knowledge placed a premium on proximity among individuals and organizations. Much of the knowledge that companies accumulated needed to be personally carried forward across technology generations by engineers and operators acting as individuals and in teams. Individuals could put their personal knowledge to work in groups without making it explicit. Knowledge that could not be put into words could be transmitted from one person to another by demonstration, and in workplace and social situations, much explicit knowledge flowed by word of mouth. These interactions had both cooperative and competitive aspects. Engineers working on similar projects in rival companies telephoned each other regularly to compare progress, spurring each other on in the race to finish first and best. Sometimes they also compared notes on problems. Managers and scientists from companies engaged in joint activities could take up long residencies at partner firms from around the world to work on projects or to help customers settle in new tools and processes. Yet in many cases the physical proximity of industry locations allowed them to return home to their families in the evenings.

U.S. companies achieved key positions in the move to high-volume production by listening to managers who challenged home-country-centric

S CREATION / 7

preconceptions of innovation, new business creation, and management processes. These managers—based both at home and in Japan—advocated business models that challenged their companies' traditional strategic orientations. They recognized that collaboration within a network of producers, suppliers, and alliance partners in a small geographic area—in this case the Kansai-Tokyo corridor in Japan—played a critical role in the knowledge creation processes that fueled industry progress. They persuaded their companies to give unprecedented power to their Japanese affiliates to take initiatives, create new capabilities, and globally manage those initiatives and capabilities from Japan. None of these companies had ever set up global headquarters for any of their businesses outside of their home countries. But each headquartered FPD operations in Japan, leveraging global technology and marketing competencies with their Japanese affiliates' technology and organizational capabilities. Each of these companies—including IBM, Applied Materials, and Corning—brushed aside the sovereignty of central management vision in international operations, just as trends toward market globalization seemed poised to reinforce it. In doing so they uncovered new principles for competing in the knowledge-driven, global manufacturing industries of the future.

## FPD INDUSTRY CREATION AS KNOWLEDGE-DRIVEN COMPETITION

Competition in emerging, knowledge-driven manufacturing industries catches companies in a double bind, demanding a continuous stream of innovations at the same time as prices seem destined to decline without end. Companies face pressures to differentiate their products from those of competitors while at the same time cutting costs. Yet any technological advantages and performance improvements they achieve through costly R&D investments diffuse rapidly. Even in the face of constant innovation, markets seem geared to instantly reduce advanced technologies to the condition of featureless commodities, distinguishable only by small price differences. In new industries' earliest generational transitions, companies must learn quickly, try to recoup their costs, and move on, despite uncertain profits.

As examples of knowledge-driven industries, new economy businesses such as software development spring more readily to mind than manufacturing.

Once developers design an innovative piece of software, a company can manufacture and distribute the program at marginal cost approaching zero. Rapid market feedback from users through such low-cost communications modes as email helps software developers to constantly improve their products. They can issue a seemingly endless succession of incremental versions. Companies can offer updates to existing users of prior generation products through inexpensive distribution mechanisms such as websites. Margins improve as average costs decline. Development, production, and distribution mechanisms wrap themselves around the globe with little if any regard for national borders.

Knowledge-driven competition presents similar management challenges in manufacturing industries, but investments in capital equipment place different limits on managers' flexibility to respond. Innovation in manufacturing industries takes place on two fronts: new product development and production process engineering.[4] New product development refers to R&D efforts aimed at bringing a product from concept to prototype. Production process engineering refers to the development and integration of manufacturing equipment and materials, as well as recipes used in manufacturing. The distinction between process and product innovation is important for understanding how new industries evolve from competition among alternative designs for implementing a new technology or functionality to competition among alternative product offerings once a dominant design has been established. Drawing on the work of W.J. Abernathy and J.M. Utterback, Giovanni Dosi, and David Teece, we refer to these two competitive phases as pre-paradigmatic and paradigmatic. In pre-paradigmatic competition, companies typically use standardized manufacturing equipment, in order to retain flexibility to adopt an alternative should their offering fail to establish itself as the dominant design. In the paradigmatic phase of competition, companies face reduced uncertainty over product design, gravitate toward customized manufacturing equipment and compete on the basis of scale, learning, and process innovation to reduce costs.[5]

Knowledge-driven competition, in manufacturing as in software, has blurred the distinctions between product and process innovation as well as between pre-paradigmatic and paradigmatic phases of competition. This has important implications for the ways in which knowledge-driven markets and

**Table 1.1** Product- vs. Knowledge-Driven Competitive Orientations

| Attribute | Product-Driven Orientation | Knowledge-Driven Orientation |
|---|---|---|
| Value Creation | cost/differentiation | speed |
| Gaining Advantage | protect, exploit, adapt | create, share, transcend |
| Sustaining Advantage | vertical integration | access and participation |
| Internationalization | efficiency/market-seeking | knowledge-seeking |
| Globalization | project, protect national positions | leverage unique national strengths |
| Nationality | isolate | collaborate |

industries evolve. Successful new products generally provide high profitability in the early phases of their product life cycles, trading on their uniqueness. But all products eventually encounter price competition as imitations enter the market. When price competition sets in, companies turn to process innovation to find means to lower costs to preserve margins. But if price competition takes hold before a technology settles down to a dominant design, innovating companies face difficulties recovering their development costs over successive generations. Companies face immediate, relentless pressures to reduce costs, while investments in R&D and updating of plant, equipment, materials, and processes continue to rise. Such conditions typify the early phases of an new industry's evolution, as companies work to simultaneously drive product innovation forward *and* cost structures down to profitably reach the broadest possible market.

### New Value Creation Logic

In this context, the true basis for competition shifts from traditional product positioning to knowledge-driven strategies, and the logic of value creation shifts from cost leadership and differentiation to speed (see Table 1.1). Companies find themselves in a race to create knowledge that finds embodiment either as product innovation or process creation, learning, and codification. Profit streams depend on the capability to drive costs down while simultaneously remaining ahead of rapidly evolving user preferences with improved products.

In a new, knowledge-driven industry, the rapidity of change abbreviates the moment of differentiation advantage companies can derive from any one instance of product innovation, while pressure to improve process, as well as to bring up new manufacturing lines to satisfy demand for the leading-edge products, remains continuous. In a seeming inversion of the laws of supply and demand, customers at times face shortfalls in their preferred product allocations at the same time as prices are falling. This occurs because the convergence of product and process innovation causes new sources of production efficiencies and new generations of product to arrive simultaneously on the scene.

The apparent failure of traditional logics creates the need to search for new ways of thinking. Sharp's Hiroshi Take regarded the experience of working on large-format color LCD commercialization from the perspective of Buddhist practice. "We have to commute between extremes," he said, "and then the middle will be open to us. We must exhaust all possible courses of action."[6]

We will see in this book how companies reconciled seemingly conflicting forces through continual knowledge creation in the period of FPD industry takeoff. As FPD producers sought to establish large-format color LCDs in the market, competition to lower costs and to improve and differentiate the product took hold simultaneously, with a vengeance. Companies engaged in R&D to continue the stream of product innovations that helped to differentiate their products. Continuous process innovation played a key role in creating a response to both differentiation and cost pressures.

The 14-inch prototypes represented a significant product innovation that demonstrated the near-term feasibility of creating an FPD that would rival traditional home TV technology—the cathode ray tube (CRT)—in performance. But before any FPD technology could replace the CRT, pricing would need to be established that users would regard as competitive. Proponents such as IBM's Web Howard were confident that users would pay a premium price for the added functionality of a very high-quality color display combined with portability in a notebook computer, a new application in which FPDs did not substitute for CRTs.[7] But establishing a high-volume FPD industry would still require cost economies to achieve pricing consistent with mass market penetration. Companies faced the dual process engineering challenge of raising manufacturing yields to a high level and achieving economies of scale to drive costs down.

Impressive as the 14-inch prototypes were, they did not establish the underlying thin-film-transistor (TFT) LCD technology as a standard. Development continued in a number of alternative FPD technologies,[8] some of which users may eventually prefer to TFT LCDs in important applications, including the large-screen, high-definition TV-on-the-wall. The demonstrations did not establish a dominant design for TFT LCDs, either. The products continued to evolve on multiple dimensions, including size, video response rates, color subtlety, viewing angle, power consumption, and weight, just to name a few. The key to overcoming many of the technical performance limitations on TFT LCDs proved in many instances to reside in the accumulation of manufacturing process knowledge, as well as its embodiment in successive generations of manufacturing equipment and materials.

The 14-inch TFT LCD demonstrations sounded the starting bell in the knowledge race to create high-volume manufacturing processes for large-format, color FPD production. In the course of creating the prototypes, the companies established low-volume R&D production lines. Establishing these lines enabled the companies to identify potential high-volume technologies for the various production stages, as well as equipment suppliers and materials makers, such as Applied Materials and Corning, who might partner with them to create these new manufacturing capabilities. In order to continue to increase screen sizes and production efficiencies at the same time, the industry needed to introduce new generations of production equipment capable of processing larger substrates and new formulations of materials that could withstand the shifting characteristics, such as high temperatures and the harsh chemicals used in the manufacturing process.

Each new generation of manufacturing facilities (known as fabs) substantially raised the ante for capital equipment. Second-generation TFT LCD fabs cost in the neighborhood of $200 million for equipment. Generation 3 fabs cost more than $500 million. Generation 4 fabs were projected to come on line sometime around 2000 at a price tag of more than $1 billion. Close collaboration among FPD producers, users, and equipment suppliers was needed to enable coordinated adaptation to generational shifts. Even so, rapid shifts from generation to generation challenged equipment and materials suppliers to recoup their development costs before each new generation supplanted its predecessor. The rapid shifts also sent the FPD producers repeatedly back to square one in settling down new manufacturing lines and raising their yields to commercial levels. Speed in bringing these new lines up to commercial yield played the essential role in determining which among the companies would gain a profit from generational shifts and which might be driven further underwater. Old lines could sometimes produce the newer, larger sizes that consumers seemed always to prefer, once introduced. But the old equipment would produce fewer of the larger display cells per substrate, at far higher costs per cell.

### New Logic for Gaining and Sustaining Competitive Advantage

Knowledge-driven competition chips away at the old logic of vertical integration, which leads companies to prefer to own all stages of a production process right down to assembly and marketing of products sold in final goods markets. In the race to create new knowledge and learn, companies seek close relationships to gain access to specialized capabilities that they can leverage in combination with their own. These include both direct and indirect relationships with competitors and potential competitors, as well as relationships with suppliers and customers. Critical bodies of new knowledge and technology accumulate in shared rather than proprietary domains of activity, such as interactions with equipment and materials makers whose customer base includes multiple competitors in the same industry. Companies that elect to excessively limit such interactions often "pay a price in loneliness," as several managers in the FPD business put it, slowing and or even halting their own technological progress. Toru Shima, president of the IBM-Toshiba manufacturing alliance Display Technologies Inc. (DTI), put it this way: "We must ask our vendors to keep reasonable security. But, if we do not disclose our problems, then time to improvement is reduced. If you are a front-runner, then some of your improvements must spill over. We view it as a kind of tax."[9]

Companies also raise up their own competitors, licensing product and process recipes to other companies that can then act either as second sources for current technology, or as continuing sources for past technologies that have moved down market. In this way, companies create, share, and transcend new technologies, remaining on the leading edge of knowledge creation while at the same time sustaining profit streams from past achievements. In the FPD industry, companies manufacturing in Japan were already aggressively licensing STN technologies and first-generation TFT LCD technologies in Taiwan and Korea even before Generation 3 production lines were fully functional at high yields. TFT LCD production process recipes, as well as the availability of up-to-date manufacturing equipment and materials, enabled companies to establish strong global leadership positions from Korea by the late 1990s. Furthermore, although almost every FPD supplier marketed notebook computers that incorporated its screens, companies rarely pursued strategies of exclusive self-supply, but also acted as merchants and customers to other manufacturers.

Most mobile computer divisions of FPD producers began early to supplement internally produced supplies with displays bought from outside suppliers that competed with them in downstream markets. FPD divisions also sold their displays to these same competitors for incorporation into their products. Most industry members, as Toshiba's Tsuyoshi Kawanishi explained, decided that the value of the knowledge that their companies could gain from buying and selling TFT LCDs outweighed the losses implied by making their best products available to competitors. As market sellers, each would be forced to meet competitors' standards at the component level, not just in products incorporating TFT LCDs, such as notebooks. Furthermore, if product divisions remained free to outsource FPDs, internal production sources would be held to external competitive standards. Merchant sales also helped to build volume to hasten the learning process needed to bring new manufacturing lines up to commercial yields and production scale efficiencies. The latter factor was particularly important during the period when companies wrestled with fluctuating yields on the first high-volume lines. The process of ramping up yields proved erratic and defied projections, with sudden success stories carrying a mixed blessing of production gluts.[10]

### New Logic for Internationalization and Globalization

Knowledge-driven competition is international competition. But it is also international collaboration. Knowledge generally diffuses more rapidly within countries than across international boundaries for many reasons, including language differences, institutional differences, denseness of interpersonal networks, and not-invented-here syndrome. Furthermore, it is difficult to transmit much explicit knowledge, and impossible to transmit tacit knowledge, where physical distance intervenes. Because of the delays inherent in international diffusion of knowledge, competitive advantage in knowledge-driven industries depends on real-time physical presence, physical access, and physical participation in national networks where knowledge accumulates first. Real-time presence, access, and participation depend critically on how companies build, manage, and globally network their local capabilities. Instead of creating and shepherding new businesses close to home, companies need to leverage affiliates' unique strengths and capabilities on the ground in countries where new industries emerge. Local strengths and capabilities include not only those housed within affiliate organizations, but also those inherent in affiliates' local networks of suppliers, customers, and alliance partners.

Many companies will find that this means transforming relationships among international affiliates from hierarchy to collegiality, as well as creating wider scope for affiliates to take the lead in managing nationally based alliances, partnerships, and supplier relationships. The U.S. companies that played the most active roles in FPD industry creation showed remarkable consistency in their approach to this issue. Applied Materials and IBM built on long-standing traditions of cultivating management teams and organizations in Japan that enjoyed global authority and accountability within specific areas of competence. Corning was transforming its Japanese organization to take a more active role in customer liaison on technical matters, as well as in R&D at about the time the FPD industry opportunity was identified. Each of these companies ultimately established the global headquarters and nerve center for its FPD-related business in Japan.

## A MAP OF THE INDUSTRY

In Figure 1.1, we have mapped the FPD industry of the 1990s, and its customers and suppliers on a matrix to show how the rapid pace of knowledge and market evolution created global interdependencies that defined the industry's strategic challenges. The vertical axis represents companies' opportunities to leverage market access and knowledge of consumer preferences.

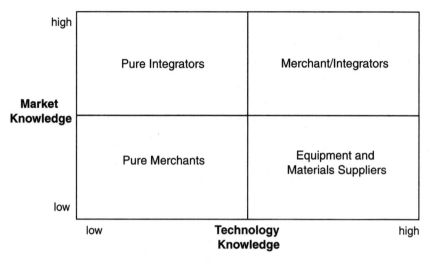

**Figure 1.1** Knowledge interdependencies in the FPD industry.

The horizontal axis represents companies' opportunities to leverage access to the knowledge-creation process in FPD technology. Companies gained these opportunities in varying degrees and combinations as FPD manufacturers, sellers, customers, and suppliers of manufacturing equipment and materials. The four categories of companies identified in the figure—*pure integrators, pure merchants, equipment and materials suppliers,* and *merchant/integrators*—represent all of the functional combinations that emerged. We define and describe each of these company types in the text that follows. Clearly some combinations offered greater opportunities for competitive advantage than others. In every case, companies' capabilities to exploit their opportunities depended on thinking that transcended traditional limitations of corporate and national boundaries on managerial decision-making.

*Pure integrators* acted as customers, assembling or purchasing assembly services to combine FPDs along with other electronic components for applications such as notebook computers (screens) or video cameras (viewfinders). This category included the U.S.-based personal computer marketing powerhouses Compaq and Dell, as well as many smaller OEM assemblers in Taiwan and elsewhere in Asia. The U.S. companies fully exploited their market access opportunities by developing capabilities in marketing, logistics, and sophisticated low- or zero-inventory materials management, but not in manufacturing per se. As successful brand-name marketers, they held the pulse of consumer preferences in the huge U.S. market, but did not function as technology innovators. Their access to leading-edge FPD components derived both from large order sizes and capabilities to feed market knowledge back to FPD producers.

*Equipment and materials suppliers* manufactured the equipment and materials necessary to produce FPDs. Most leveraged knowledge gained manufacturing semiconductor fabrication equipment, optics, or chemicals. In this way they made critical contributions to the early knowledge base of the new FPD industry. Semiconductor equipment makers, optics suppliers, and many FPD producers shared a history in the integrated circuit (IC) and DRAM industries. Building on shared knowledge and past relationships, many suppliers with semiconductor roots played key roles in the knowledge creation processes that made possible the industry's rapid advance following the transition to high volume.

Industry members demarcate the generational phases of technology advance according to the size of the substrates, or motherglass used as raw inputs. The rapid growth of the Internet caused consumer preferences for larger personal computer display sizes—particularly the most advanced, large-format, color TFT LCDs—to evolve at a pace that caught the industry by surprise. Producers needed to rapidly increase substrate sizes to respond to these preferences, gain efficiency, and raise capacity. Many equipment makers were used to the comparatively gradual size increases of semiconductor wafers.

The ability and pace of change in the race to enlarge substrate sizes depended on the rate of advance in the manufacturing equipment and process technologies deployed in the most advanced fabs. Since the first advanced display fabs opened in 1987, most industry participants agree that the technology advanced at least four generations, called "Gen 0" through "Gen 3," each with its variants. Some variants sufficiently departed from their root generations to warrant the claim of fractional "Gen dot.X" status. (In the wake of Gen 3, innovating producers have claimed Gen 4 or even Gen 5 status for variants that others refer to as fractional. We will avoid this controversy and use the fractional terminology.) Counting all of the declared dot.Xs, eight generations passed between 1987 and 1998 (see Table 1.2). This rate of progress compelled equipment and materials suppliers, system integrators,

**Table 1.2** Main Commercial Generations of Color TFT LCD Substrates

| Generation | Typical Substrate Size | Optimized for Display Size (qty.) | Earliest Adoption: Startup Dates, Adopters |
|---|---|---|---|
| 0 | 270 × 200 mm | 8.4-inch (2) | 1987 Sharp |
| 1 | 300 × 350 mm | 9.4-inch (2) | 3rd Quarter, 1990 NEC |
| | 300 × 400 mm | 10.4-inch (2) | 2nd Quarter, 1991 DTI |
| | 320 × 400 mm | 8.4-inch (4) | 3rd Quarter, 1991 Sharp |
| 2 | 360 × 465 mm | 9.5-inch (4) | 2nd Quarter, 1994 Sharp |
| 2a | 360 × 465 mm | 10.4-inch (4) | 2nd Quarter, 1994 DTI |
| 2.5 | 400 × 500 mm | 11.3-inch (4) | 3rd Quarter, 1995 Sharp |
| 3 | 550 × 650 mm | 12.1-inch (6) | 3rd and 4th Quarters, 1995 Sharp, DTI |
| 3.25 | 600 × 720 mm | 13.3-inch (6) | 1st Quarter, 1998 Samsung |
| 3.5 | 650 × 830 mm | 17.0-inch (4) | 3rd Quarter, 1997 Hitachi |
| | | 15.0-inch (6) | |
| 3.7 | 730 × 920 mm | 14.1-inch (9) | 3rd Quarter, 2000 Samsung |

SOURCES: Authors' analysis based on comparison of business press, company documents, and interview materials.

and FPD manufacturers to work together continuously to optimize new processing capabilities with the shifting substrate dimensions. FPD producers and equipment and materials suppliers labored jointly to install and integrate the multiple pieces of equipment in each new line. They remained in intimate interaction throughout the painful ramp-up phases, working together to run the new lines at full capacity for months while struggling to bring yields of usable product from near zero to commercial levels of 80–90 percent. The knowledge that accumulated in these interactive processes began immediately to inform the design of subsequent generations already on the drawing board.

Rapid generation changes imposed particularly heavy costs on equipment manufacturers and some materials suppliers. Before they could recoup investments made in current generations of equipment and materials, many faced the need to introduce new products to accommodate new substrate sizes. Due to their lack of direct access to consumer markets, equipment and materials makers faced difficulties securing shares of the financial returns from these transitions, despite playing central roles in the necessary technological advances.

These uncertainties, the relatively small size of the emerging industry, and significant resource requirements deterred many potential equipment and materials suppliers with relevant competencies from getting into the business. Many U.S. companies hedged by signing on to government subsidy programs that provided incentives to form learning partnerships for new projects with the few small-volume FPD producers and development companies that manufactured in the United States. Most of these companies found it difficult or impossible to translate limited, home-based experiences into sales to high-volume manufacturers.

Senior managers of U.S. companies that accepted the risks and entered the global high-volume industry envisioned its potential as a platform for innovations that would replace or influence the state-of-the-art in their core businesses. Corning's entry into glass substrates seemed logical, in part because the company had decided to transition away from its long-standing position as a leading supplier of glass "bulbs" for television picture tubes. Applied Materials managers justified the risk of entering equipment manufacturing as a way of learning to work with the large substrates used for high-volume FPD fabrication. Their investment anticipated increases in the size of silicon substrates used in Applied's long-standing semiconductor equipment manufacturing business. We will see how differences in mindsets regarding knowledge creation and cross-national learning partnerships made the critical difference between success and failure for companies in this sector.

*Merchant/integrators* refers to companies that manufactured FPDs, sold them in world markets, purchased FPDs from others, incorporated FPDs along with other electronic components in branded products such as notebook computers, and in some cases made some of their own manufacturing equipment and materials. As sellers of branded final goods incorporating

FPDs, merchant integrators could take advantage of direct feedback from market segments they chose to address. As FPD manufacturers, merchant/ integrators participated with equipment and materials suppliers in creating the knowledge that fueled the industry's technological progress. As a network of suppliers and customers to one another, merchant/integrators comprised a forum in which members shared their perceptions of consumer preferences, product performance trends, market dynamics, and technology trajectories. External sourcing helped merchant/integrators' notebook divisions stay abreast of market trends, pricing trends, and technological possibilities. Competing with outside suppliers to meet the requirements of both internal and external customers forced FPD manufacturing divisions to remain at the cutting edge in technology, costs, and product features. The product flows that the companies exchanged embodied the technology and market knowledge that each accumulated.

Evidence suggests that merchant/integrator strategies addressed high-potential opportunities. During most of the 1990s, all of the top FPD producers and two of the top four notebook producers were merchant/ integrators. These included Sharp Electronics, the IBM/Toshiba alliance DTI, NEC and Hitachi in FPDs, and Toshiba and IBM (independently) in notebooks. In fact, conventional wisdom in the U.S. during the early 1990s suggested that the Japanese merchant/integrators could unfairly control world competition in FPDs as well as in products, such as notebooks, that incorporated them. But little credible evidence ever accumulated to support this view.

Much credible evidence accumulated that calls it into question. IBM's success offered an early, obvious contradiction. As DTI pushed out the limits of display sizes in the 1990s, parent firms IBM and Toshiba did not preempt other notebook makers from supplies of the most advanced displays, even when their own notebooks were back-ordered. Both continued to buy and sell displays on the open market, regarding connections to other producers and customers as learning conduits too vital to disrupt. Long-term commitments also needed to be honored. Both Toshiba and IBM derived advantages in the global notebook market from their joint ownership of DTI's FPD manufacturing facilities. But the advantages grew out of early design-in opportunities for new notebook screens, rather than from restricting customers' access to DTI's output. IBM's and Toshiba's decisions to act as merchant/integrators rather than treat DTI as a vertically integrated captive source is emblematic of a major distinction between product- and knowledge-driven strategies. Product-driven strategies sustain advantages by excluding competitors from access. Knowledge-driven strategies create future advantages by sustaining learning relationships.

Few U.S. industry participants learned much from the IBM example, however, while many dismissed DTI as a Toshiba dependency. Then new merchant/integrators began emerging in Korea. Since 1999, several Korean companies have taken places among the leading FPD and notebook producers. As this book went to press, Samsung and LG.Philips vied over the number one and two global market share positions in FPDs. Several large system integrators from outside of Asia have also taken ownership positions and created alliances with firms that owned FPD manufacturing capabilities. Philips leveraged existing European FPD production capabilities by entering alliances with Hosiden in 1997 and LG in 1999. Later in 1999, Apple aligned itself with Samsung by making an equity investment.

These developments highlight the prescience of the IBM and Toshiba managers who established the joint research projects in the mid-1980s that led to the DTI alliance. Both companies proceeded despite ambiguity about future markets for products that might incorporate displays. Managers in both organizations envisioned FPDs as a critical learning project necessary to renew competencies and sustain advantages as their traditional markets transformed. But during the period from the mid-1980s and mid-1990s, many other U.S. and European companies with similar capabilities—for example, AT&T, Xerox, and Siemens—faced the same ambiguous dynamics and apparent needs for display functionalities in their downstream product markets. How and why did IBM alone forge ahead to help found the high-volume industry? We will examine how management and decision-making processes differed between companies that forged ahead and companies that held back.

*Pure merchant* companies fall into the remaining cell of Figure 1.1. By the end of 1999, every pure merchant FPD company that employed mainstream technology had either combined with a systems integrator or left the industry. The casualties included Japan's Hosiden, which joined with Philips to create the Hosiden and Philips Display Corp., and OIS, a small subsidiary of Guardian Industries in the U.S., which closed its doors. Other small companies that existed independently or as venture projects on the peripheries of larger organizations—the Xerox FPD company dpiX, and the ImageQuest subsidiary of Hyundai, for example—also consolidated with customers or closed.

Contrasted with the successful entry of new FPD merchant/integrators, the failure of any pure merchant to survive the 1990s in the high-volume industry says something about inherent pitfalls in fast-evolving, knowledge-driven industries. Throughout the industry's early history, technology and manufacturing process milestones resulted from knowledge accumulation across companies, continuity of learning within companies, and intercompany collaboration to create new knowledge. The most successful companies combined the roles of customer and supplier, connected first-hand to consumer markets, and collaborated closely with equipment and materials suppliers whose offerings encompassed the industry's state-of-the-art.

Most of the merchant firms' FPD businesses were headed by individuals who had achieved important display research milestones and enjoyed industry-wide reputations as technology leaders, entrepreneurs, or both. Yet the lack of direct connections to consumers and demand-side experience in FPDs limited these companies' learning opportunities and cash flows. Many U.S. merchants also restricted their interactions with world-class equipment and materials suppliers due to capital constraints or by imposing national origin criteria in their supplier selection processes. We believe that despite strong technology bases and individual leadership, the merchant companies fell out of step with the state-of-the-art in high-volume manufacturing because they did not participate fully with other industry sectors in the knowledge-creation process.

## KNOWLEDGE INTERDEPENDENCE AND THE INEVITABILITY OF COOPERATION

We have argued that the FPD industry's rapid advance resulted from a continuous, cumulative knowledge-creation process. Jared Diamond has pointed out that all technology development is cumulative, and that the individuals whom society credits with great inventions have generally improved on preexisting versions of these inventions originated by others. Most uses for technologies are found after they are invented, he argued, and society recognizes great inventions as such only when it is capable of finding a widespread use for them.[11]

The FPD experience was consistent with this observation. The current leading technology for high-information-content FPDs emerged from basic research programs that unfolded mainly in the United States prior to 1973. RCA scientists made critical advances in building LCDs: displays that combine a light source, electronics, and liquid crystal to create a flat visual medium. Research at Westinghouse established methods to deliver electronic impulses that convey an image to a screen by using thin film transistors (TFTs) that can be coated by the millions onto a display's rear glass substrate. By acting as tiny switches powered by row and column drivers on a display's edges, TFTs provide the means of addressing individual pixels, or microscopic picture elements, to display and refresh an image at sufficient rates of speed so that the human eye perceives motion. This means of pixel addressing is known as active matrix. (We discuss mainstream and alternative FPD technologies in greater detail in Chapter 3.) The high-information-content FPD technology commercialized in today's high-volume, large-format FPD fabs is called TFT AMLCD, for thin film transistor active matrix liquid crystal display, or TFT LCD for short.

In the late 1960s and early 1970s, U.S. companies encountered difficulties finding product applications that consumers would accept for the new flat displays. The dream of flat, wall-hanging television primarily motivated FPD research programs. Once researchers had solved some of the basic problems, an affordable, reasonably sized flat TV seemed too futuristic to warrant immediate investments toward commercialization. Yet in several Japanese companies, researchers who learned of the U.S. research programs saw immediate uses for the new technologies. The applications they had in mind appeared small and rudimentary compared to the ambitious products that U.S. firms envisioned. Their relatively modest objectives, however, functioned as achievable goals that brought FPD commercialization within the realm of short-term possibility. FPDs emerged from the laboratory. Society had found a use for them.

The earliest commercial FPD manufacturing facilities were established to serve two Japanese companies' supply requirements for small, branded consumer products. In the early 1970s, as part of their respective strategies to improve their calculators and watches, Sharp Electronics and Seiko Suwa separately pushed ahead with R&D programs that led to the commercialization of TFTs. Many observers credit the subsequent product introductions as the first steps toward the foundation of the high-volume industry.

Both Sharp and Seiko began early to seek outside customers for their displays. The companies found that engagement with outside customers diversified and invigorated their R&D efforts. The companies continued to upgrade their FPD technology to meet their own future product needs and to meet customers' needs and specifications. Apple Computer, in particular, acted as an early, influential Sharp customer. Apple entered both the notebook and personal digital assistant markets early. Its mouse- and icon-driven operating system preceded Microsoft Windows by a number of years. The visual demands of Apple's operating system added great impetus to FPD producers' early quests for color, high resolution, size, and smooth video motion.

As the industry's potential grew increasingly evident in the mid-1980s up to the notebook computer's takeoff in the early 1990s, joint efforts among manufacturers, equipment suppliers and materials manufacturers were needed to enable the transition to high-volume production of the largest, most advanced displays. Sharp was again involved, along with Toshiba and IBM, as FPD producers, and Applied Materials and Corning in equipment and materials. Several of these companies competed with each other in related fields. Cooperative relationships with downstream system integrators such as Apple Computer and Compaq continued to play a role, even as some FPD producers began to compete with them for shares of the notebook market. Without these cooperative relationships, the high-volume FPD industry would have emerged eventually, but not when it did.

## WHY INDUSTRY STUDIES? A NOTE ABOUT METHODOLOGY, FINDINGS, AND RELEVANCE

In the 1990s, it became fashionable to suggest that the knowledge basis of competition in new, information-based industries differed fundamentally from the product basis of competition in old, manufacturing-based industries. Our experience in the capital-intensive FPD industry indicates that knowledge-driven competition has changed the basis of success in manufacturing industries as well. Knowledge creation and learning capabilities increasingly separate winners from losers in all industries. Understanding new industries born to this new order offers a chance to distill and grasp new fundamentals.

On the other hand, why jump to conclusions? Didn't the Internet asset bubble of 1999–2000 demonstrate the danger of generalizing principles too quickly from short-term phenomena? Other eras in business history have brought fundamental transitions that were not widely appreciated for their impact on practice until many years later.[12] Why not wait to see which new companies and business models stand the test of time?

We find this view shortsighted, if not dangerous. The escalating pace of competition, technology evolution, and globalization lends a new urgency to the study of emerging industries. These factors take on added weight in new manufacturing industries with their demands for commitments to increasingly costly fixed assets. Lessons not drawn quickly lose their relevance. Learning may come too late when a company has already saddled itself with inefficient manufacturing lines, as some FPD companies did repeatedly as substrate sizes increased.

Lessons drawn more quickly by competitors can render a company irrelevant. Widely accepted evidence suggests that early events in an industry's history, such as IBM's choice of Microsoft DOS as the operating system for the PC, have an ever-increasing power to lock in a future course of industry evolution and a distribution of returns among participants.[13] Following the early histories of new industries can help us understand the common themes within strategies that put companies in the paths of early opportunities when they arise. Understanding the knowledge basis of these paths and opportunities may also help companies to remain flexible while making the minimum investments necessary to exploit them.

Viewed from a U.S. perspective, certain themes stand out starkly in the FPD industry's history. Once high-volume manufacturing emerged in Japan, U.S. notebook suppliers moved rapidly to establish their primary sourcing relationships there. Most U.S. merchants, merchant/integrators, and manufacturing equipment and materials suppliers arrayed themselves in two groups. The first established authority and accountability for their FPD businesses in Japan, or in the case of some equipment and materials suppliers, focused heavily on serving Asian customers. The second group of companies established U.S.-centric strategies that interfered with their abilities to learn. They remained aloof from contacts in Japan, reluctant to commit funds to commercial-scale production and often relatively isolated from each other. Many immersed their personnel and capital in drawn-out R&D projects to look for alternatives to the mainstream TFT technology.

FPD business performance differed widely between these two self-selected groups of U.S. companies. The internationalized group—all businesses started by major U.S. corporations—established leadership positions in their product areas. The domestically oriented group, which included businesses started by major U.S. corporations, small corporations, and entrepreneurial startups, experienced a high rate of attrition.

The performance differences between these two groups effectively presented us with a natural experiment. We found that we could isolate a number of common strategy and organizational factors that varied strongly and consistently between the high- and low-performing groups. We could also see important differences within the groups in strategies and organizational processes. We examined the strategies of FPD companies that originated outside the U.S. to see how these same common factors might apply. In many instances, we found that successful companies from other countries shared similar orientations. This discovery increased our confidence in the validity of the factors as critical dimensions of strategy for any company operating in the industry. The common factors form the basis of the general ideas that we offer as necessary but not sufficient for successful strategy in emerging high-technology industries. The differences define the company strategies that created the unique competitive dynamic of the global FPD industry and U.S. companies' roles in it from inception to the year 2001.

In the mid-1990s, large Korean companies created the second wave of firms from outside of Japan to begin high-volume FPD production. We believe that most Korean companies, in addition to exploiting their proximity to Japan, fundamentally shared the successful U.S. firms' orientations. Important differences arose due to distrust arising from the political heritage of a colonial relationship between Japan and Korea, and the availability of tools, materials technology, and process recipes from the United States, Europe, and Japan.

## OUR RESEARCH CHANGED OUR THINKING

The research for this book began in the United States. We started following the FPD industry's evolution in 1990, just after the first large-scale plants opened in Japan. Our mental maps of the industry resembled those of many other U.S. industry participants and observers. We viewed the industry as a fragmented, politically constructed set of competing national fiefdoms.[14] During 1996 and 1997, thanks to a project grant from the Alfred P. Sloan Foundation's Industry Studies Program, we visited more than 160 company and government officials interested in the industry, some repeatedly in the years after, to develop and test our thinking. Our travels took us to every major producer that existed in Japan, Korea, Taiwan, Europe, and the U.S. at the end of the industry's takeoff phase during the late 1990s. We also visited most minor producers, many major equipment manufacturers, materials suppliers, and key customers ("users"). This fieldwork allowed us to discover for ourselves the limitations of creating knowledge from a distance. We became convinced that overcoming these limitations resides at the core of successful business strategies not only in the FPD industry, but in any emerging industry.

We set out in our research to understand the U.S. industry experience. We found that it was impossible to understand the experience of a U.S. FPD industry, because no such industry has ever existed. We could understand only the global industry and U.S. companies' experiences in it. This book describes our experience in the global FPD industry and our efforts to translate that experience into practical ideas for managers, consultants, academics, government policy-makers, and others interested in building successful strategies for managing in the early years of new, high-technology industries. These ideas owe much to the managers who participated in our study, who shared with us their reflections on the day-to-day realities of creating new businesses in a world where countries have boundaries, but technologies do not.

# 2

# What's Wrong with This Picture?

By the year 2010, most households in industrialized countries will own at least one thin, cineramically dimensioned, wall-hanging video screen that will reproduce super-realistic images from high-definition television signals. Automobiles will typically offer global positioning information and virtual streetscape navigation aids on mid-dashboard screens. Clunky CRT computer monitors will have almost entirely disappeared from office and home desktops in favor of sleek easel- or wall-mounted flat panel displays. Flat panel display technology will have revolutionized transportation instrumentation from buses to jet airliners to space shuttles, because it will offer superior ruggedness, space economies, visual fidelity, non-radiance, operate over a broad temperature range and consume relatively little power. FPDs will pervade and change almost every province of life, including health care, home appliances, telecommunications, information storage and retrieval, scientific instrumentation, recreation, apparel, visual arts . . . multimedia will be everywhere. By 2005, according to the market research firm DisplaySearch, the business of making FPD components to assemble in all of these applications could top $70 billion per year.[1]

In 1994, examples of products fitting all of these applications and more already existed. Sales of displays reached $8.4 billion, an increase of 45 percent over 1993.[2] In early 1996, a leading market observer projected that by 2001, sales would reach $19 billion.[3] Influential technology and corporate strategy thinkers wondered publicly, in best-selling books and in countless corporate boardrooms whether any firm with a hardware stake in the digital revolution could prosper in future markets without FPD manufacturing capabilities. They intended this as a rhetorical question.

In the U.S. in the early 1990s, many managers, scientists, media members, academic experts, consultants, and government officials believed that as FPDs approached commercialization, U.S. firms were late for the second act and stood to miss a great deal more of the play. In 1993, nearly 95 percent of all FPD manufacturing took place in Japan (more than 99 percent of the most advanced displays), even though in U.S. dollar terms, U.S. notebook computer makers accounted for a large, rapidly growing proportion of consumption. By 1995, as much as 5 percent of advanced display capacity had migrated to Korea. Korea—respected U.S. high-tech business and public policy analysts assured anyone who would listen—had set a national priority to establish FPD manufacturing competence. Capacity was growing in Asia, but U.S.-located capacity was not keeping pace. Europe appeared to be falling even farther behind. The question on the lips of the U.S. experts was, can a *country* hold on to a stake in the digital revolution without FPD manufacturing capabilities?

The question evidenced a strange national myopia. Several U.S. multinational companies—including IBM, Corning, and Applied Materials—played central roles pushing FPD technology past the brink of commercialization in the late 1980s into high-volume manufacturing of large-format color displays in the early 1990s. They have retained and enhanced their leadership positions ever since. But the U.S. business press did not celebrate the managers responsible. During the period when IBM, Applied Materials, and Corning built their businesses, several other prominent U.S. companies, including AT&T and TI, received more attention when they abandoned the industry. Yet in spring of 1994, the Clinton administration announced a major subsidy campaign to bring U.S. companies into the business. Like prophets through the ages, the leading U.S. FPD industry participants found themselves without honor in their own country.[4] What did managers in the successful U.S. companies know that had escaped the notice of government policy makers and other U.S. companies demanding government help? How had the successful companies escaped notice?

## CLUSTER BUSTING

Many managers and public policy-makers believe that when a scale-intensive, high-tech industry concentrates in one country, companies from other countries get easily locked out. Debates about how countries should respond to high-tech industry concentration in other countries have centered on either building countervailing industry concentrations, known as clusters, at home or establishing facilities within the foreign cluster itself. Public policy-makers and business strategists[5] have turned for guidance to economic geography, which offers a research tradition that explains why certain industries develop great centers of creativity and productivity in particular world regions but not in others. Attention has focused on country- or region-specific management or innovation systems, path-dependent historical developments, institutions such as great universities and national research laboratories; on knowledge spillovers that occur among companies through common suppliers, consultants, customers, job changers; and on the social and professional networks that emerge as part of the local industry community.[6]

The FPD experience demonstrated how easily these ideas can be misappropriated as guides to corporate strategy and public policy, particularly in the early days of a new high-technology industry. Much U.S. thinking about the FPD industry has foundered on the notion that the vitality of the FPD industry in Japan somehow arose when factors intrinsic to Japan combined (illegitimately!) with a U.S. invention. The proposed factors ranged widely across well-known Japanese business institutions and country capabilities, including the availability of patient capital, the coordinative power of government, and the meticulous rigor of Japanese engineers and production workers. None of these factors can explain the collision of individual creativity with the global innovation system that catalyzed the beginnings of LCD research in Japan.

Cluster thinking confused many observers of the FPD industry's emergence, because it draws attention to stable internal institutions and knowledge that may offer countries some degree of autonomy in world markets. The Northern Italian high-fashion textile industry, for example, may well have enjoyed an ability to dictate important trends in high-end fabric design for a time. But such autonomy is increasingly short-lived. Even traditional industries like high-quality fabric have diffused to Asia in recent years due to globalization. Focusing on clusters creates a false sense of permanence for strategy.

More important, high-technology industries increasingly emerge from a convergence of local with global factors and knowledge that catalyzes rapid accumulation of new knowledge. The high-volume FPD industry originated in a convergence of knowledge drawn from a variety of countries. The industry's concentration in Japan in its early phases was a consequence—not a cause—of the rapid acceleration of knowledge accumulation around FPD technology in the 1980s. As the mass consumer market for notebook computers emerged in the 1990s, industry learning continued to be catalyzed by global forces, including the Internet and fast-changing consumer markets for technology products that incorporate FPDs.

State-of-the-art business strategy prescriptions for entering the FPD industry in the early 1990s would have suggested establishing operations in Japan. But potential market entrants that waited until the FPD industry's strength in Japan had become widely evident were already too late to play leadership roles in that phase of the industry's development. Leadership was important, because only leaders made any money. As a consequence of the financial stress many companies in Japan experienced, it later became possible to buy in using an acquisition strategy. Only one company, Philips, was wise enough to do so. The foundations for Philips' decision were established years earlier in its little-known close affiliation with Matsushita and its own efforts to establish FPD manufacturing in Europe. In general, if a company discovers the attractiveness of an industry because a cluster has emerged somewhere, its management has already experienced a fatal failure of foresight.

U.S. public policy prescriptions for the FPD industry focused on finding government-led means to remedy the U.S. market's presumed failure to offer incentives for local firms to establish facilities on U.S. soil. In economic theory, market failure offers one of few justifications for government intervention in markets. International markets can fail for many reasons. Knowledge markets are especially prone to failure because one firm's ownership of knowledge does not preclude other firms from having it, whether or not they pay for it. Firms face difficulties in negotiating knowledge exchanges: price-setting by nature involves some degree of disclosure, and disclosure of information reduces incentives to pay.[7] Particularly in the U.S., these difficulties have predisposed managers to focus on strategies that restrict outsiders' access to their firms' knowledge, rather than on ways of profitably sharing what they know with competitors, collaborators, suppliers, and customers. These concerns intensify for most companies when they manage international businesses.

The heated pace of high-technology competition inverted this conventional logic for some U.S. firms. As a consequence, they became key players in the FPD industry. But the U.S. government fell behind by implementing policies to encourage domestic FPD industry cooperation in preference to international market activity. These efforts to create a countervailing FPD cluster in the U.S. created incentives for U.S. companies to cut themselves off from the suppliers, customers, complementary assets, and knowledge streams that were creating the industry. U.S. taxpayers and some entrepreneurs in the FPD industry paid a heavy price for these failed policies in the 1990s. Instead of establishing high-volume FPD manufacturing in the U.S., another generation of progress was lost (see Chapter 8).

Five years of intensive research on the evolution of the global FPD industry has persuaded us that high-tech industry concentration in one country or world region does not lock out companies from elsewhere unless they close the door on themselves. New high-technology industries often bubble under the surface for many years in several countries before they suddenly achieve critical mass and commercialize at global scale in one or more of them. Once a new industry emerges, continuity in knowledge accumulation, the pace of technical advance, and the commercial and social relationships that drive knowledge creation reinforce one another.

It is impossible to predict the exact timing and location in the world where any given technology will commercialize and a global industry emerge. But it is possible for companies to design management processes that positively affect their probabilities of taking part. Companies with affiliates in a country or region where an industry emerges have as good a shot as local companies at taking integral positions, provided their managers can fully leverage local organizational capabilities with global technological capabilities as these opportunities arise. In the successful companies in our study, local managers functioned in peer networks as global managers. Local initiatives served as primary means to identify and go after global opportunities. Long-standing corporate research traditions in underlying technologies combined with strong local operations to establish these companies' stakes in the rapid accumulation of knowledge assets associated with the FPD industry's emergence. Developing such a knowledge stake formed a necessary condition for successful physical asset deployments anywhere in the world, including at home.

## GENERALS FIGHTING THE PREVIOUS WAR

Many experts we respect predicted dire consequences for U.S. national competitiveness if U.S. firms failed to establish FPD manufacturing capabilities on U.S. soil. Many argued that FPDs represented one of a handful of keystone technologies that a dominant manufacturing country can use to threaten other countries with a stranglehold on future competitiveness. Others hinted that the geographic concentration of FPD production represented "yet another example"[8] or even "an outstanding example"[9] of questionable international competitive practices orchestrated under the Japanese government's strategic guidance. FPD technology, after all, originated in the U.S. So did many other technologies in which Japan at times seemed destined to establish global manufacturing dominance, including semiconductors and television. In these perspectives, FPDs represented a classic example either of a market crippled by unfair trading practices and government intervention, an injury to national competitiveness due to U.S. corporate smugness and short-termism, or both.

The analogy to the semiconductor trade wars of the mid-1980s to mid-1990s seemed apt. In the 1990s, the semiconductor industry remained an enduring emblem of technology trade tensions between the U.S. and Japan. Until the mid-1980s, more semiconductors had always been produced in the U.S. than in every other country in the world combined. When Japan took the lead, many U.S. companies, media observers, and government officials searched for explanations in Japanese industrial policy first and in U.S. corporate strategies a distant second.

FPDs have a lot in common with semiconductors. They are technically complex electronic components subject to cyclical prices. They require transcendently expensive, large capacity, ultra-clean fabrication facilities to achieve commercial yields and economies of scale. In the mid-1990s, one FPD fab cost over $500 million for capital equipment alone and as much again to ramp up to industry standard yield rates. In 2001, companies expected to spend at least twice as much for a fab. Corporate strategists and observers of the Japan/U.S. relationship on both sides of the Pacific drew an analogy between the two industries, and it stuck.

The semiconductor trade dispute created a legacy of self-defeating conventional wisdoms for U.S. companies about high-technology competition with Japanese companies. The U.S. lacked patient capital to assume large risks for the long haul. Capital costs in Japan were lower. U.S. workers lacked the discipline to achieve high yields in tedious, claustrophobic cleanroom manufacturing environments. Furthermore, process engineers were in greater supply and better trained in Asia. FPDs were commodities with infinitely declining prices and margins. Only Asian companies, which put survival above profit, would accept the low margins that persist in a commodity manufacturing market. Furthermore, the products would remain in chronic oversupply and have little impact on competitiveness in downstream markets. The U.S. was too far behind in the current FPD technology paradigm to catch up with Japan's lead. Better to remain patient and work toward a leapfrog. The best technology will win in the long run.

Conventional wisdoms imply conventional solutions. Why not concentrate on gaining U.S. government financial support for intra-industry cooperation to build up the FPD equipment and materials industry within the U.S.? This had worked in semiconductors. Hadn't it?

Defense analysts amplified the case for government intervention by raising issues of military preparedness. FPDs were supplanting old-fashioned CRT video screens in aircraft and would play an important role in equipping foot soldiers, weapons, and transport on the battlefields of the future.[10] U.S. security, these analysts argued, depended on creating a domestic FPD manufacturing source. Furthermore, the source should operate at commercial scale to avoid the high costs associated with cultivating small-scale, captive military suppliers. The FPD equivalents of $600 coffeepots were to be avoided.

In spring 1994, the Clinton administration responded by bundling a number of existing FPD development programs operated by the Defense Advanced Research Projects Agency (DARPA) together with several new manufacturing promotion initiatives to create the National Flat Panel Display Initiative (NFPDI).[11] The White House proposed funding of over $500 million for the U.S. industry to match in a crash program to catch up with the industry in Japan by creating high-volume FPD manufacturing under U.S. ownership on U.S. soil.

## THE HIGH-VOLUME U.S. FPD INDUSTRY: MISSING IN ACTION

The U.S. crusade to create a domestic FPD manufacturing base obscured the reality of a global industry in which U.S. companies had already established significant leadership positions. By 1996, U.S. multinational companies, working through alliances and affiliates, could claim the leading global market shares in the dominant FPD technology and application (31 percent of active matrix LCD displays for high-end notebooks) as well as shares ranging from 60 to 80 percent in key FPD equipment and materials categories. In 1997, these companies—including IBM (TFT LCDs), Applied Materials (CVD), Corning (substrates), 3M (brightness enhancement film), and Photon Dynamics (test equipment)—continued to lead in global markets while gaining significant technical ground. During the same period when many U.S. corporate officials, public policy-makers, educators, consultants, and media critics prominently voiced concern about the lack of U.S. competitiveness in FPDs, the first three of these companies had already positioned themselves as indispensable players in the global industry's take-off. They subsequently led the industry in a number of critical transitions and breakthroughs.

Yet within the FPD and computer industries, the basis, nature, and scope of U.S. successes remained widely misunderstood, controversial, and misrepresented. Critics explained U.S. FPD industry developments in a manner consistent with assumptions drawn from the semiconductor experience. In doing so, they framed U.S. companies' successes as the nation's failures. These companies, critics suggested, had moved manufacturing offshore to take advantage of Japan's low cost of capital, disciplined workforce, abundant process engineers, subsidies offered through the Japanese government's Ministry of International Trade and Industry (MITI), and to avoid funding research into leapfrog technologies that might supplant the dominant TFT LCD technology. The ventures represented opportunistic components sourcing or technology borrowing operations that allowed Japanese partners, suppliers, and customers to hold U.S. companies hostage to their superior capabilities. At best, critics framed these ventures as U.S. corporate sellouts. More typically the ventures were ignored. U.S., European, and Asian consultants, market researchers, academics, and media, for example, conventionally credited the market share of DTI, the IBM/Toshiba alliance, to Toshiba.[12]

Realism in understanding the global FPD industry from a U.S. perspective begins by framing U.S. global players in, not out, and focusing on success rather than failure. Even a small sampling of the facts we encountered in our research raises questions about the assumptions that typified the debate about U.S. companies' participation. We learned that Corning and 3M manufactured mainly in the United States, not Japan. Applied Materials' joint venture with Komatsu, Applied Komatsu Technologies (AKT), centered its manufacturing and R&D in the United States.[13] The successful U.S. companies provided technology, financial, and marketing leadership in every case, and they continue to do so. Many critical R&D and operational issues for DTI were addressed at IBM's Thomas Watson Lab in Yorktown Heights, New York.

Each of these companies achieved a leading market share by entering the business early with contributions that removed technical roadblocks and enabled industry development. Skilled engineers interested in FPDs have found themselves in oversupply rather than the other way around, while opportunities in Japan and Korea have drawn expatriates homeward. For example, TFT LCD pioneer Shinji Morozumi, who in the mid-1990s served as deputy general manager with primary responsibility for FPD manufacturing at Hosiden's TFT LCD Business Division, worked for a time on FPDs at Xerox in Silicon Valley. Process engineers are easier to hire in the U.S. than in Japan, where most large companies exact lifetime allegiance from key employees. Land, water, and cleanroom operators all come more cheaply in the U.S. than in Japan. MITI has not strategized the industry in Japan. "If you can figure out how to do it, please let me know," commented Soichi Nagamatsu, the MITI official responsible for tracking the industry.[14] He had met with industry indifference, if not resistance, in his main project to encourage cooperation to stabilize continuously changing glass substrate dimensions long enough to reach a standard. "Things are changing too quickly. The industry is not mature. Companies are looking for advantages by jockeying for larger screen sizes."

## KNOWLEDGE BEGINS AT HOME . . .

The story of the FPD industry is a story of knowledge accumulation by companies and countries. But even more than that, it is a story of individuals working together and in solitude to create that knowledge. As Ikujiro Nonaka and Hirotaka Takeuchi have pointed out,[15] knowledge begins with the individual. Much knowledge remains tacit. Knowledge that remains tacit ultimately cannot be verbalized, written, or explicitly represented in any codified manner. This does not mean that it cannot be transformed in part into explicit knowledge, shared in part, or recombined with other knowledge to create and accumulate new knowledge. The ability to share and recombine tacit knowledge, as well as explicit knowledge that remains verbal but not otherwise codified, depends in part on direct observation, interpersonal interaction, and demonstration, which in turn depends on proximity to events and other people. So does the ability to learn from experience. These processes reside at the core of technological innovation and commercialization.

In fact, for many processes of knowledge creation and sharing, tacit and explicit, but strictly verbal knowledge make the difference between success and failure. Successful FPD companies recognized this in formulating plans to start up new production lines. Companies generally assigned many of the same engineers and operators as those who started up previous lines to "wake up" new lines, especially when the lines incorporated new generation technology. One leading company lost market share and profitability when it stretched its complement of experienced engineers and operators over several startup lines at the same time. All of the startup lines fell behind schedule. Yields also fell on the company's existing lines because too many experienced staff members had been reassigned.

Knowledge creation draws energy and context from the realm of shared experiences. Individuals may share experiences with as few as one other person. But many varieties of experience can be shared among individuals that comprise teams, companies, alliances, or other relationships among companies, professions, communities, industries, countries, or beyond. Individuals who share experiences do not necessarily interpret the experiences in the same ways. For that matter, they may not share the same interpretation of information, either. Shared place can have an impact on interpretation and enhance processes of mutual sense-making and agreement. During the mid-1990s, for example, companies whose managers directly experienced and participated in the emergence of the FPD industry in Japan tended to adopt optimistic demand forecasts in their planning processes. Companies that tied their fortunes strictly to U.S. or European facilities tended to adopt less optimistic sales forecasts, overestimate capacity, and misinterpret the implications of technology changes for capacity utilization and industry output. As a consequence, they presented less promising business cases to the financial community, and at times even perceived production gluts when, in fact, shortages existed.[16] The two groups also used different consultants and market research firms. Japanese companies and those with experience in Japan generally relied on Japanese statistics and forecasts by Nomura. Nomura's forecasts generally offered a more optimistic picture of industry prospects than did the forecasts of other companies. For example, during the early 1990s, Nomura held to year 2000 sales forecasts of around $15.6 billion, while the U.S.-based Stanford Resources Inc. forecast that sales would range somewhere above $8.5 billion.[17]

Proximity in itself, however, guarantees nothing. Organizational processes are needed to enable individuals to share and recombine both tacit and explicit knowledge. Successful companies in the FPD industry overcame process barriers to knowledge creation within their organizations across hierarchy, organizational functions, businesses, and countries. They overcame some of these same barriers and others in their interactions with companies that were their suppliers, customers, collaborators, and competitors. Talented individual researchers and managers have presided over amazing achievements within the organizations of both successful and unsuccessful FPD market entrants. Companies from around the world that succeeded in FPDs took visionary gambles, some of a "bet-the-company" magnitude. The differences between membership in a winning team or a losing team depended partly on luck. But the odds in the game depended heavily on the existence, nature, and continuity of knowledge creation processes that connected individuals and teams within companies, among companies, within countries, and internationally.

None of these individual types of connections were sufficient in themselves to create success. Many display producers who restricted their operations to U.S. territory, for example, remained technologically abreast or even ahead of basic research developments in Japan at the same time as their commercialization efforts began to feel the negative effects of isolation. Many U.S.-based FPD researchers enjoyed relatively better international connections to engineering and science colleagues around the world than to business decision-makers within their own firms, at potential U.S. alliance partners, and within the U.S. financial community. The U.S. development company Plasmaco, for example, played a leading role in a particularly promising approach to commercializing Plasma Display Panels (PDPs), which are widely believed to represent the future of very large, wall-hanging, digital, high-definition home TV. Yet founder, president, and chief technology officer Larry Weber consistently failed in efforts to raise capital in the U.S. to move toward production. Matsushita engineers viewed one of Weber's prototypes and met with him at the annual conference and exhibition sponsored by the prestigious, international professional group, the Society for Information Display (SID). The engineers immediately recognized the promise in Plasmaco's approach and brought it to the attention of top managers in Matsushita's Consumer Electronics and TV businesses. Weber soon met with Matsushita Electric Industrial Company Corporate President Yoichi Morishita, who agreed to provide project financing, barely rescuing Plasmaco from bankruptcy. In January 1996, Plasmaco became a wholly owned Matsushita subsidiary. Weber remains at the helm as president and CEO.[18]

## YOU HAD TO BE THERE . . . BUT IT COULD HAVE BEEN ANYWHERE

The FPD industry, like the high-tech industries in Silicon Valley, appears to have thrived on the mutual geographic proximity of its members. But it is important to emphasize that the industry's early concentration in Japan arose as a consequence of knowledge accumulation driving the technology forward and not the other way around. There was nothing intrinsic about Japan, Japanese management style, or any other Japanese business, academic, or government institution that uniquely destined Japan to serve as the global center of the industry. Sharp and Seiko established a foundation by translating basic research into a stream of products whose success stimulated the continuity of knowledge accumulation within these companies. But just as the stimulus for the original research originated outside the boundaries of these companies and Japan, the pace of knowledge accumulation soon demanded new global outreach if it was to be sustained. The outreach required direct connectivity among companies and people because of the creative nature of the task at hand.

Applied Materials, for example, could not meet Sharp and Toshiba's early demands for new FPD manufacturing solutions simply by taking blueprints off of the shelf. Applied needed to leverage its semiconductor-based core competencies with those of Sharp and Toshiba to create new knowledge and apply it in the new area. Core competencies cannot be separated from organizations, groups, and individuals. The factors driving the need for direct connectivity operated at each of these levels of the companies' operations. At the organizational level, the new knowledge was transforming each company's competencies as applied in its semiconductor as well as FPD businesses. For the companies' knowledge and competencies to co-evolve in this fast-paced learning environment, intimate, continuous interaction needed to happen. At the individual and group levels, fast-paced learning overtaxes processes that transform personal (or tacit) knowledge into explicit knowledge and explicit knowledge into knowledge that is written down or otherwise represented in replicable fashion. Continued progress relies increasingly on individual and group interaction to incorporate the full breadth of new knowledge in successive phases of learning. In these ways, increasing speed of knowledge creation establishes powerful forces for companies in the same new industry to co-locate their operations.

Corning, Applied Materials, and IBM built their mastery of the FPD business on a foundation of physical science plus physical presence. Each of these firms established organizational capabilities and customer relationships in Japan many years prior to the FPD industry's take-off in the early 1990s. These capabilities and relationships made possible timely, face-to-face knowledge exchanges among engineers, scientists, and product planners in Japan and the U.S. concerning final goods, equipment and materials market opportunities, manufacturing process challenges, and key vectors of technological advance. Affiliates in Japan did not behave passively as corporate eyes, ears, and implementers, but actively as strategists, advocates, and builders of the FPD business. The companies divided responsibility and authority on a non-hierarchic basis between U.S. and Japanese operations for product planning, R&D, manufacturing, and marketing decision-making. Japanese and U.S. operations divided leadership roles in the businesses depending on distinctive country-based organizational competencies and market opportunities. They worked together with the other elements within their firms' global networks, each playing the role of leader and follower simultaneously.

The most successful participants acted early to form cross-national alliances and supplier/customer relationships to advance materials and manufacturing process technology. These advances enabled the industry to take off when it did and the companies to create sustainable leadership positions in it. Managers at Corning, for example, encountered their first clue about the potential of the company's fusion glass to produce defect-free, inexpensive, ultrathin, ultraflat substrates for FPDs when unsolicited laboratory orders arrived at the company's Japanese affiliate. Corning's technology eventually drove down FPD producers' glass costs more than 90 percent.[19] Applied Materials' new chemical vapor deposition (CVD) equipment pioneered large-area thin film deposition techniques that replaced a far more primitive technology based on solar cell manufacturing. Industry standard yields improved as a consequence from between 50 and 60 percent to above 90 percent of throughput. IBM and Toshiba formed Display Technologies Inc. (DTI) as an outgrowth of a joint research project led in the mid-1980s by IBM Japan. DTI established a position at the center of a network of equipment and materials suppliers leading the way in creating process technology to produce larger screen sizes.

Managers of companies without mature ties to Japan's markets lacked firsthand knowledge of the beginnings of the advanced FPD industry and did not believe the potential existed for its subsequent explosive rate of market growth. They miscalculated the rate of technological improvement in the dominant TFT LCD technology, underestimated demand, overestimated supply, and adhered to hierarchic organizational structures and decision-making rules that devalued input from their affiliates in Japan. They adopted self-defeating analogies to other industries in which U.S. companies had faced tough competition with Japanese companies. These factors led to several attempts to "leapfrog" TFT LCD technology with approaches that remain peripheral to overall industry advance. "Buy American" equipment and materials policies, not-invented-here syndrome, and autarkic partnership and acquisitions strategies kept many U.S. companies from acquiring the experience and technology necessary to improve the efficiency of U.S. fabs. The domestic politics surrounding many public funding programs also curtailed interaction between participating U.S. companies and the U.S. and Japanese companies that should have been their best suppliers, customers, and development partners. The proponents of a territorially based U.S. FPD industry increasingly peripheralized themselves.

### WILD DUCKS, COOL SCIENCE, AND ANTICHRISTS

The contrasting experiences of the IBM/Toshiba alliance versus AT&T and its unsuccessful alliance negotiations offer portraits of how intra-company, inter-company, national, and international knowledge-creation processes affected FPD commercialization decisions.

Early in Toshiba's deliberations about entering TFT LCD manufacturing, for example, Tsuyoshi Kawanishi contacted a number of potential partners and suppliers including Applied Materials and IBM. Discussions met a receptive ear with IBM Japan senior executive vice president Nobuo Mii, an old friend.[20] IBM was in the process of rethinking its display commitments. The company had nearly decided to back away from PDP FPD technology because it seemed likely to suit TV better than computer applications. But under the leadership of IBM executive vice president Jim McGroddy, the commitment remained strong to maintain continuity in FPD research at the Thomas Watson Laboratories in Yorktown Heights, New York. With the shift away from PDPs, top managers at IBM Japan saw an opportunity to propose a global initiative in developing TFT LCDs. They put together a team, pressed for a mandate, and secured it in collaboration with Yorktown Heights.

IBM and Toshiba began joint research at Toshiba facilities on August 1, 1986, to develop an approach to TFT LCD technology and manufacturing process. During the first year, IBMJ senior technical staff member for Display Technology and project leader Shinichi Hirano supervised the building of state-of-the-art cleanroom facilities at IBM. The second year of the project continued there. Steve Depp, director in the IBM research area responsible for the FPD project, recalled, "When we first traveled to Japan to work on the R&D line, each of us paired off with a partner to run one piece of glass through the line. Every partner was a specialist in a particular process step. Despite the language barrier, we came to a very quick mutual understanding of the challenges. It helped us get connected to one another."[21] The projects proceeded with a policy that paired off IBM and Toshiba people for every experiment.

At the end of the second year, having created a 14.26-inch prototype ("still working today," Hirano said in late 1996), both companies were positioned to undertake manufacturing independently, but it was not clear that IBM would move into production. Yorktown Heights researchers brought the prototype into an executive conference where "abstract strategic issues" and financial models had predominated in an ongoing debate between "wild ducks," who wanted to plunge ahead with display production, and "antichrists," who had treated TFT LCDs' potential for IBM with skepticism all along. Display project member Bob Wisnieff recalled, "People can see things logically before a demo. Once they see (the demo), they feel it emotionally. . . . once the display was lit, it was 'we have to do this.' They wanted the project."[22]

Lucent Technologies, on the other hand, shut down its Bell Laboratories FPD R&D manufacturing line in March 1996. Lucent, the recently spun-off former equipment division of AT&T, was restructuring its research activities to focus on projects that related directly to current markets in which the company offered products. Lucent's FPD project members included respected figures who had spent many years in the industry, such as former IBM wild duck Web Howard, who had joined in 1992, not long after DTI got its first plant up and running. At the time of the shutdown he was serving as president of the SID.

When we visited headquarters in New Jersey on April 11, 1996, canvas Lucent Technologies banners were tied down with ropes over the AT&T logo on the buildings and grounds. In the entrance atrium, management was presiding over an all-employee event to mark the company's initial public stock offering, which took place one week earlier, and at $3 billion,[23] set a record as the largest in U.S. history. Upstairs, Web Howard was preparing to take some personal time before joining a new venture, FED Corp., one of several U.S. startups developing an upstart FPD technology.

Over the previous two years with surprising speed, the thirty AT&T display project members had awakened a fully articulated, scalable, operational TFT LCD fabrication line and used it to design a manufacturing process. In the cleanroom down the corridor, the project's several remaining members were dismantling the remaining $20 million worth of manufacturing equipment, packing most of it up to be shipped to Ann Arbor, Michigan. There it would be reassembled and used in research and education at the University of Michigan's Center for Display Technology and Manufacturing. Because no former AT&T project members accompanied the gear, however, much of the process knowledge that went with the equipment was doomed to be lost forever.

Although managers depicted the project as an inevitable victim of the Lucent spinoff/restructuring, the corporate and industry isolation that killed it appeared to have set in much earlier. Linkages to AT&T businesses and top managers were weak. Product divisions, focused on current markets, were looking for immediate bottom line contributions, not costly associations with research projects. According to one Lucent manager involved in the shutdown decision, top management knew little about the program. As he put it, "When you are doing a risky project, you want to wait until you have a success before you apprise top management."

Highly centralized and focused on the U.S. market, AT&T lacked organizational resources in Japan. Two successive alliance negotiations conducted by the FPD project management—one in Japan and one in Korea—were terminated on the brink of finalization by the next higher level of AT&T management. The AT&T representatives had enjoyed considerable leverage in the negotiations. Their Asian counterparts had expected to be able to demand money and markets in return for bringing a less-competent U.S. partner up to speed in technology and manufacturing. The AT&T reps surprised them by demonstrating technology mastery and quick design of an extremely robust manufacturing process. In both cases the negotiation failures left the prospective Asian partners rueful, facing less optimistic targets than were planned for the alliance. In one instance, AT&T was forced to honor a compensation clause in the negotiations framework. The Asians pressed ahead on their own and eventually established themselves as industry leaders.

The negotiations failed because AT&T product groups declined to make the financial commitments necessary to bring the project to fruition. Ultimately, as another executive close to the project put it, "They lined everyone up, shot 'em and then poured salt on the earth." All but one of the thirty team members left the company. A manager in another U.S. firm commented,

"The biggest waste was, when they do need displays—and they will need them sooner rather than later—they won't have retained any of what was learned. They won't even have the expertise left in-house to outsource."

## * INDUSTRY AS COMMUNITY

Shinichi Hirano enjoyed comparing the global community of FPD researchers, engineers, and managers to an imaginary, small, Midwestern American town in the late Victorian era. "*Winesburg, Ohio,* not Silicon Valley," he said.[24] The analogy to the town Sherwood Anderson portrayed in his famous book of character sketches[25] hits home in more ways than Hirano probably intended. As in Winesburg, Ohio, in the flat panel display industry almost everyone knows everyone else. There's quite a bit of talk, but a lot of information travels from person to person despite remaining unspoken. Secrets have a way of getting out, and a certain amount of rumor and disinformation also makes the rounds. There are heroes, saints, and rebels but also jealousies, victims, and disappointments. The industry has its own ways. People rarely seem to leave. Failure is forgiven, success sometimes met with antipathy. The industry is not very large . . . yet. In 1999, each one of the top ninety-one firms in the *Fortune* 500 enjoyed higher revenues than all of the FPD producers in the world combined. The industry is young but has already faced adjustment to a radical economic change with the coming of the digital age, hypercompetition, and the Asian Economic Crisis, just as the Midwestern and southern U.S. faced the industrial revolution following the American Civil War. And as in the book that inspired Anderson to write *Winesburg, Ohio,* Edgar Lee Masters' *Spoon River Anthology,* many who have populated the industry are remembered mainly by their epitaphs.[26]

Unlike Winesburg, Ohio, in the 1890s, however, the FPD industry swiftly changed throughout the 1990s. Participants have called the changes exponential, inexorable, out-of-control, unnecessary, self-destructive, unsustainable, dangerous. Several suggested that the evolutionary pace more than doubled—or even quintupled—the developmental pace of semiconductors during the analogous period in that industry's history.

Companies that succeeded in FPDs navigated the fast-paced years of the industry's early development by implementing strategies to accumulate knowledge assets *wherever* the opportunities arose. Openness to knowledge creation and sharing was the root underlying cause of companies' successes. Location of the industry in Japan was the consequence of the timing and speed of knowledge accumulation that occasioned its takeoff. As the industry continued to evolve, companies that focused on knowledge assets acquired product and process technology that provided them with flexibility to locate production anywhere in the world which, in principle, could have included Winesburg, Ohio. Companies that focused on physical assets diminished their options and have either changed their strategies or disappeared.

# 3

## Continuity Under Adversity: Wild Ducks, Cool Science, and Antichrists

*For thirty years we could say we would have wall-hanging televisions in ten years. Now, I can say, 'next year.' The only reason I'm here is that I hung in there. U.S. Corporate America will not hang in there for thirty years or even twenty . . . maybe for ten.*

LARRY WEBER[1]

*Truly I say to you, no prophet is acceptable in his own country.*

LUKE 4:24[2]

No one knows what the Old Testament prophets really looked like. But Larry Weber could probably play one on screen. In fact, on one of his own screens. He has the experience. In April 1994 in a San Jose, California, friend's garage, Larry Weber sprinted the final, breakthrough lap that made his tiny company a contender in the relay race to commercialize the full-color, alternating current plasma display panel (AC PDP). AC PDP has since emerged as the leading candidate technology for the largest of the wall-hanging screens many observers expect to dominate the high-definition home television market of the near future. But on that day, Weber pulled his company just a step or two ahead of bank foreclosure.

Larry Weber co-founded Plasmaco in 1987. At the time he was a University of Illinois research professor heading the world's first and perhaps leading PDP research lab. As Plasmaco's chief science officer, he dedicated the new development/production company to advancing the PDP technology invented at Illinois by his predecessor and dissertation advisor, Donald Bitzer. Weber located a home for the new venture in an old cider mill. He spotted the mill while flying his airplane around upstate New York, keeping an eye out for empty parking lots. Weber equipped the facility with new cleanrooms and eighty-eight truckloads of used R&D and production gear from the closed IBM PDP fab down the road in Kingston, New York.

Weber tangled right from the start with partners, investors, and board members who saw little hope of success in his vision to move the company away from monochrome PDPs toward a color future. By 1993, the company's monochrome displays were selling for less than cost. At the end of the summer, a new CEO specialized in turn-arounds resigned after six weeks. He regarded the situation as hopeless. Plasmaco's board added the CEO title to Weber's chief scientist responsibilities (to his consternation). Then they walked away and left the company for dead. Cash resources dwindled. Most staffers remaining on the job accepted pay reflecting shortened work weeks. But many put in full-time hours anyway, loyal in adversity to a man they regarded as a patriarch of the flat television dream. After all, Weber's former Illinois research colleagues, fellow alums, and Ph.D. students worked in and led most corporate and university PDP research programs around the world, as well as programs investigating other FPD technologies. How could he not succeed?

The good news was that no one any longer paid attention to whether Weber used company resources to work on monochrome, color, or anything else for that matter. CEO Weber promptly cancelled monochrome production, conserving the last of Plasmaco's modest stockpile of cash, raised some money privately, dipped into personal savings to pay for materials, and relaunched his AC PDP color research program. Five months later, desperate to unveil the new technology at the 1994 Annual Meeting of the SID, Weber loaded the prototype and himself onto the last plane he could catch for San Jose.

The new display was not quite finished. Weber planned to program the driver circuitry using wave form equations published in an engineering journal by a group of scientists doing PDP R&D for a competitor. Weber rushed from the plane to the convention exhibition floor, where he joined scores of scientists, engineers, technicians, executives, exhibit architects, and artisans preparing furiously for the meeting's opening. The cacophony of hammers, machine tools, and audio merchandising programs rang in his ears as he prepared to throw the switch that would light the display for the first time. He threw the switch.

The picture frame remained utterly inert.

Something was wrong with the equations!

Weber removed the equipment from the exhibition floor and secluded himself at the home of a nearby friend where he worked non-stop throughout the meeting's four days and four nights. He tried everything, anything to get the display to work. With less than three hours left of the meeting, it flickered to life. Back on the exhibit floor for the closing two hours prior to adjournment, Larry Weber's full-color AC PDP created the sensation of the meeting. "We were so short of funds that we had to hard-wire a test-pattern for the demo," he later recalled. One of Plasmaco's bankers had postponed foreclosing on the company until 5:00 P.M. Tuesday afternoon following the meeting. "We had the last of the payment ready to bail us out at 4:45 P.M." But things would get worse before they got better. After nine more months of dramatic technical progress but hand-to-mouth financing, Larry Weber and his remaining partners were looking for an angel. In January 1996, they agreed for Plasmaco to be entirely acquired by Matsushita Electronic Industries of Japan, famous in the U.S. for the Panasonic brand name. Larry Weber remained as president and PDP *sensei* to scores of researchers in Matsushita's laboratories around the world.[3] By the year 2000, Plasmaco designs formed the core and leading edge of Matsushita's impressive line of plasma flat panel TVs, attaining an industry-leading 60 inches of diagonal screen size in a prototype shown at the SID annual meeting. The Society also awarded Larry its prestigious Karl Ferdinand Braun prize, named for the inventor of the cathode ray tube (CRT), in recognition of his "pioneering contributions to Plasma Display Technology and its commercialization."[4]

## THE DISTANT VISION

The dawn glare of the information age has obscured the television age as an epithet for contemporary times. Displays of electronic images and information from new media saturate daily life. Yet television remains the most pervasive mass communications and entertainment medium in human history. More to the point for our purposes, the conventional color TV monitor, based on Karl Braun's 1897 invention, has remained the workhorse of human eye/machine interfacing for new and old media alike. And why not? CRTs offer relatively low manufacturing costs and display warm, realistic moving images at wide viewing angles. In the grand scheme of things, alternative display technologies only recently started to approach CRTs in these critical areas of user appeal.

## CRTs: The Benchmark

All video-capable display technologies portray action by relying on some method of electronically depicting and then rapidly refreshing a changing series of images. Because CRTs refresh the images so rapidly (generally sixty times per second), the human eye perceives change as a continuous moving image rather than a series of distinct frames. CRTs use an electron scanning gun (cathode) positioned at the back of a glass vacuum "bulb" to bombard phosphors coated on the front glass plate (anode) with electrons. The electron bombardment activates the phosphors so that they emit light. The cathode paints and refreshes the video image by scanning rapidly in lines from the top to the bottom of the anode.

Yet visions of a holy grail, a "TV on a wall," to replace the CRT have beckoned engineers and entrepreneurs since the beginnings of broadcast television in the 1930s. Long before the possibilities of mass television became real, the idea of transmitting moving images for wall-size reproduction entered the popular imagination in science fiction novels and films. The reality of the tiny images emitted by the early, radio-set-sized televisions of the late 1940s must have simultaneously amazed and disappointed many viewers. Engineers faced an immediate challenge to increase picture quality and viewable screen size while somehow reconciling TV monitors' growing volume and weight to real-life living spaces.

This challenge has grown rather than diminished with time. The typical early twenty-first century 36-inch diagonal screen size CRT, for example, weighed in the neighborhood of 160 pounds and required a dedicated cabinet depth of 30 inches or more for room placement. Bumped to 40 diagonal inches, the sets added another 30 or more pounds. Furthermore, CRTs' curved viewing surfaces caused pictures to distort at the edges, reducing the size of the viewable area in comparison to that of flat screens with the same physical dimensions. In operation, they generated considerable heat and consumed a great deal of power. Inspired to overcome these limitations of power, placement, mobility, and image reproduction, several generations of scientists and engineers have dedicated careers to the quest for a flat panel that can produce superior moving images of any size on a display just a few millimeters thick. In light of the quest's appeal to public and scientific imagination, its significance, and the rewards that seemed almost certain to accrue to successful champions, the Larry Weber story seems incongruous.

Yet Larry Weber's travails as a scientist and entrepreneur were not exceptional. They hold much in common with the experiences of others in both small and large companies who quested after the flat panel vision. Like others, Weber possessed the skill, creativity, and independence of spirit needed to bring the vision to reality. Also like others, Weber faced daunting challenges of business organization and strategy on the path toward realizing the dream. Many found the business challenges to creating commercially viable FPDs at least as difficult to surmount as the technology challenges. Many more projects and ventures expired of business causes than of technology shortcomings.

## THE RACE TO THE STARTING LINE

In this chapter we will examine internal organizational challenges that confronted managers and research leaders as they worked to establish a theoretically desirable functionality—dynamic flat information display—as a technological reality. The chapter deals with the management of processes that challenge any organization that strives to bring a new technology from early life to the brink of high-volume manufacturing.

Companies' initial approaches to these management issues play a vital role in their later success, because early events in an industry's evolution exert powerful influences on the path of its future development.[5] In a knowledge-driven industry, learning economies can kick in early, before many companies recognize the competitive and collaborative interdependencies with others that define the enterprise community. This occurs in part because technology advances exponentially and at an ever-increasing pace. New developments evolve from the constantly expanding foundation of previous knowledge.[6] Yet no one can predict with certainty when, where, or with whom key findings will arise, the financial value of such findings at any given moment, or the sequence of developments that will lead to a technological breakthrough. Companies must race to the starting line without knowing for certain who else might run in the race, what the value of the prize might be, or whether a race is taking place at all.

Facing up to these unresolved, random elements in the early phases of a new industry's evolution can bring out the fatalist in many managers, who focus on market, geographic, or technological outcomes that can lock in particular solutions or industry structures.[7] But the pace of change in knowledge-driven industries invests such outcomes with an increasingly transient nature. Furthermore, individual companies rarely assemble all of the knowledge needed to establish or sustain a dominant position in a new industry. Companies derive advantage through specialization, continuous knowledge creation, and judicious collaboration. Successful companies implement management processes that provide their organizations with early, continuing, formative influence, plus access to the overall knowledge-creation chain in their industry as it evolves.

The Larry Weber quote that opens this chapter crystallizes the matter. Success comes from "hanging in there." Continuity. Of course, it stands to reason that in situations that carry an element of randomness (catching a wave in surfing, for example), persistence and readiness will pay off more frequently than their alternatives. It is also true that if we were to study the development paths of a group of companies selected on the basis of their success at a given time, we should expect good odds that the paths will appear continuous in hindsight. But the record shows that the people and companies that succeeded in the FPD industry sustained their progress in the face of considerable adversity, while others abandoned good fortune and technological leads.

During the early days of a new technology, the greatest catalysts as well as obstacles to continuity most often exist in internal organizational processes

**Table 3.1** Strategy Processes for Managing Continuous Development Paths
for New Technologies

| Attribute | Product-Driven Orientation | Knowledge-Driven Orientation |
|---|---|---|
| | TECHNOLOGY PROCESS | |
| Technology Focus | Leapfrog in existing applications | Target new applications |
| Development Trajectory | Build down | Build up |
| Goal Orientation | Big wins and blockbusters | Small wins and design-ins |
| | MANAGEMENT PROCESS | |
| Top Management View | The big picture | A large canvas |
| Top Managers' Roles | Decide and rationalize | Advocate and mentor |
| Research Culture | Gamble your career | Gamble your ideas |

that bias individual mindsets and behavior to either encourage or discourage new initiatives. In many organizations with active FPD development programs, the continuing debate over how to proceed resolved into clashing perspectives very much like the conflicts the IBM TFT development team sometimes referred to as "wild ducks" versus "antichrists." We have broadened our framework comparing product- and knowledge-driven competitive orientations to capture some of the key differences in organizational processes that these confrontations reflect.

Table 3.1 outlines this comparison on two dimensions. Technology processes include strategies for motivating progress from functional conceptualization to new business creation. Managerial processes encompass key roles and orientations of senior managers, including top managers responsible for allocating resources to projects as well as managers immediately responsible for the technology development process.

Our dimensions reflect some important generalities. In many instances, wild ducks prevailed to create the foundation for a knowledge-driven strategy that ultimately led their companies to succeed in the high-volume FPD industry. In some instances, wild ducks were driven off, with their ideas, to establish new FPD ventures. Some of these ventures survived and met their promise. Others failed or fell short. Still other projects died within the companies that initiated them, their wild duck protagonists scattered to other projects, other companies, or retirement.

What created the differences? In the remainder of the chapter, we will try to answer this question by explaining the ideas in Table 3.1. We will illustrate these ideas with examples from the FPD industry's development prior to the opening of the first high-volume plants. While we will try to isolate ideas and examples, it is important to note that no single managerial process determined success. Successful companies integrated these processes, although none succeeded on every attribute of every dimension. Some companies succeeded on many attributes of one or more dimensions, but ultimately failed to make a mark by failing to integrate them.

## TECHNOLOGY PROCESS AND THE FIRST GREAT VIDEO DIVIDE

The early 1980s were times of great promise for researchers and business people interested in advanced display technologies. Professional conferences like the annual meetings of the SID were dividing into two worlds. Amid deafening noise on the show floors, increasing numbers of participants jostled down crowded aisles to view companies' exhibits of ever larger, more vivid CRTs. Removed from the crowds and racket on other floors of the convention centers and nearby hotels, workshops proliferated on technical issues for video-capable, high-information-content flat panel displays. What's more, there were demos of tiny FPDs, postage-stamp-sized prototypes. Peering closely, IBM's Bob Wisnieff recalled, attendees would ask each other in hushed tones, "Is it on?"

FPD technology had reached a crossroads. One by one over the previous decade of the 1970s, RCA, AT&T, GE, HP, and Westinghouse had abandoned FPD development programs, the last two in 1979. Before the end of the decade of the 1980s, 14-inch-plus color FPD prototypes by Sharp, an IBM/Toshiba joint research project, and later other companies would move out of the technical sessions onto the exhibit floor where viewers could directly compare them to CRTs. In 1990 and 1991, first NEC and then DTI and Sharp would open the first high-volume production lines for large-format, color TFT LCDs.[8]

All of these companies originally signed on to the quest of finding an FPD—a leapfrog—that could replace the CRT. All started out with significant technical resources in at least one of two areas of competence—CRT and integrated circuit technologies—that successful companies ultimately blended in varying proportions to create their new FPD capabilities. Most companies were serious players in both areas. But a comparative assessment of these companies' most apparent technological capabilities and business needs in 1970 would have poorly predicted their relative FPD achievements by 1990.

Most observers would have bet on RCA, AT&T, and Westinghouse. RCA and Westinghouse owned significant CRT businesses. AT&T, the only company lacking CRT experience, entertained visions of pervasive videophone networks, a pop futurists' icon at the time. In the 1970s, RCA and AT&T still owned two of the world's most prestigious corporate research laboratories. RCA scientists at the David Sarnoff Research Center[9] discovered the optical properties of liquid crystal, one of the basic materials that comprise liquid crystal displays. At AT&T's Bell Laboratories in 1947, John Bardeen, Walter Brattain, and William Shockley had invented the transistor, which replaced the vacuum tube, revolutionized electronics, and is the basic component of which integrated circuits are constructed. Thin-film transistors, today coated on glass in arrays of millions to drive the most widely-used LCD FPDs, were significantly advanced during the 1970s by a Westinghouse research team led by T. Peter Brody.

In the early 1970s, IBM reached its height as the global computing mainframe giant and also operated extensive, prestigious research laboratories. But the decentralized, personalized, display-centric computing world that began to emerge in the early 1980s—as well as the strategic crisis it sparked at IBM—remained largely unforeseen. Nonetheless, in the late 1970s, IBM was the first company to commercially market a gas plasma flat panel display. The display sold well, principally for financial data display applications in trading rooms. Toshiba and Sharp would have appeared less obvious as candidates for leadership. Toshiba, primarily a components supplier, led the vanguard in developing process technology for low-power C-MOS memory chips, and was building a second- or third-rank OEM market share position in CRTs.[10] Sharp, the first Japanese company to license television patents from RCA, was just emerging as an independent technology force. The company had long relied on components sourced from other companies—including picture tubes—to implement the innovative product ideas of its engineers. In 1970, management placed a "bet-the-company" gamble on Sharp's first semiconductor fab, funding the project in part by withdrawing from that year's Osaka World Fair.

Each company that initiated FPD research—whether it persevered or ended up relinquishing the quest—managed the development process in a different way. But there were also important common factors among the companies within each of these categories. These differences and common factors provide a number of useful, generalizable lessons for managing in the early phases of any new industry's development. We will discuss those lessons in detail in the remaining pages of this chapter, loosely following the organization of Table 3.1.

### Technology Focus at RCA, Sharp, and Seiko: Ugly Princes and Leapfrogs

Companies increase their chances of sustaining continuous development paths for new technologies by striving to reach a series of progressive, incremental goals rather than a single grand objective. Leapfrog technologies rarely spring fully mature from the lab, ready to supplant existing technologies in popular applications. In FPDs, companies that targeted wall-mounted flat TV as a sole objective ended up with glorious, high-stakes, all-or-nothing missions that management found fiscally wise to consider terminating at every major turning point. Companies that set less auspicious objectives were able to leap ahead in commercializing technologies by incorporating rudimentary FPDs in new applications that were relatively unproven in the market. Cash flows from less glamorous products such as calculators and watches kept FPD research programs solvent. These products also provided platforms for establishing manufacturability, solving manufacturing process problems, and improving display functionality to meet the needs of more demanding applications. In this way, companies gradually built the knowledge base that they would ultimately leverage to achieve their grand objectives.

RCA fell into the high-stakes category. At a May 28, 1968 press conference at its Rockefeller Center headquarters in New York City, company officials unveiled a "very crude prototype" that many of them hoped would soon transfigure into the first flat TV. This TV of the future was the first liquid crystal display presented to the general public. The tiny panel showed a black-and-white image of two moving lines.[11] The engineers demonstrated other LCD display applications, including an electronic clock that was widely shown in print and on TV news programs around the world.

The demo culminated years of Sarnoff Center research on liquid crystals, at the time a relatively obscure family of materials (see box on *Liquid Crystals*). Richard Williams had demonstrated at Sarnoff around 1960 that a liquid crystal substance in its transparent state turns opaque, scattering (or reflecting) light instead of transmitting it, when charged with an electric current.[12] Starting in 1964, George Heilmeier led some of the first experiments that harnessed this property to create an image-capable display, fueling the research program that ultimately led to RCA's announcement. Sarnoff's engineers discovered liquid crystal material that retained its crystalline properties over a wide temperature range and used transparent electrodes and a polarizer to electrically control the liquid crystal's optical properties. They called their method "dynamic scattering." With the dynamic scattering breakthrough, commercial release of the first flat television entered the range of feasibility, the company asserted in response to journalists' inquiries.[13] But many technical problems remained to be resolved before flat TV could become a reality. Most serious among these, researchers needed to find cost-effective, manufacturable means to electronically address the complex mosaic of tiny picture elements, or pixels, that would be needed to display a well-defined, moving image. The prototypes' electrodes were patterned to replicate the simple shapes required for the functionalities the research team was prepared to demonstrate.

Back in the lab after the last interviews had been granted, the Sarnoff LCD group renewed its focus on the technical and manufacturing problems that remained to be solved. As the months rolled by, the prognosis for a quick TV product introduction appeared no more immediate, but the group prototyped and sent up proposals to management for all kinds of other products: clocks, watches, simple and sophisticated instrumentation. None met with much interest. The proposed intermediate applications seemed beneath the dignity of the company that had invented commercial radio and then the television broadcasting industry. CRTs remained a healthy business for RCA. Management regarded the flat TV as "blue sky"—too far off in the future to warrant any great push.

This attitude represented a departure from RCA's historic stance. Under the entrepreneurial leadership of David Sarnoff, RCA had consistently advocated new technologies that disrupted its industries and its own products' markets. Television had supplanted radio as the dominant mass medium. Color slowly began to supplant black-and-white TV after RCA's NBC Television Network, alone and at great cost, broadcast for years in color to the tiny group of consumers who had purchased RCA's color sets. But as Sarnoff's health declined prior to his death in 1971, a new generation of managers assumed the reigns. RCA began to diversify away from consumer electronics. Management was reluctant to support a technology path that would increase costs in the short term while threatening the CRT cash cow they were milking to build their new RCA. Discouraged LCD Lab Group members began to drift away to other jobs with more immediate prospects for creating commercial products that would incorporate advanced displays. In 1973, the company made a brief foray into LCD manufacturing for point-of-purchase displays and later watches. But within a few months of starting operations in Somerville, New Jersey, RCA sold the plant to Timex, the watch company. By 1974, RCA's LCD program was dead.[14]

## Liquid Crystals

The Austrian botanist Friedrich Reinitzer discovered liquid crystals in 1888, a class of organic compounds that, in effect, exhibit two melting temperatures rather than one. Liquid crystal describes the intermediate phase in which the compounds flow freely, like liquids,[15] but exhibit a "crystalline ordering in the arrangement of their molecules."[16] The liquid crystal compounds used in modern LCDs exhibit this property over a relatively wide temperature range encompassing normal conditions of use.

LCDs employ a particular type of liquid crystal described as "nematic" or threadlike. These rod-like molecules tend to align with each other along their long axes, but remain otherwise free to move in relationship to each other. Steve Depp and Web Howard used the metaphor of a school of fish to describe the phenomenon.[17] In the next box we will explain how methods were developed to electronically control the alignment of liquid crystal molecules to create optical properties suitable for advanced displays.

In 1970, Sharp management chose FPDs along with C-MOS chips to play a central role in the company's strategy to liberate itself from dependence on components suppliers. The need to source picture tubes from outsiders for its television business had always aggravated Sharp's managers. "We had to buy CRTs, which for customers is our face," head of the TV/video group and later president Haruo Tsuji used to remind young Sharp engineers. "The display is the most decisive factor for establishing quality in the minds of consumers."[18] But why invest in old-fashioned CRT production, Sharp's managers asked each other, when RCA had pointed the way toward the video future?[19] Like RCA engineers before them, Sharp's engineers could envision a range of potential applications for LCDs. Instead of building its own CRT factory, Sharp licensed RCA's dynamic scattering LCD technology in 1970.

Unlike RCA, Sharp set out promptly to incorporate the new technology, complete with warts, into a commercial product: the hand-held calculator. The company, then known as Hayakawa Electric, pioneered the business in 1964 with the Sharp Compet, the first fully electronic calculator to be manufactured at commercial scale.[20] By the early 1970s, intense price competition had broken out as companies vied to broaden the market to include students and households, an idea once considered preposterous. Calculators had grown ever smaller since the introduction of the nearly 60-pound Compet. Engineers at Sharp decided to push the portability trend to the limits of user comfort. They soon realized that display thickness and battery size presented the main obstacles. Project leader Atsushi Asada and Isamu Washizuka, calculator division engineering section leader, assembled a multidisciplinary team of chemical, electrical, and mechanical engineers, determined to move LCD display technology out of the laboratory and onto store shelves.[21]

In April 1973, Sharp introduced the EL-805, the first calculator with an LCD display. The machine was 2.1 centimeters thick, weighed 200 grams, and burned through a set of batteries every 100 hours. The display had problems. It consumed a lot of power. Its slow response time caused calculation results to materialize like shimmering apparitions, leading a major customer to refer to the EL-805 as "the spooky calculator."[22] But Sharp had put a foot in the door of display technology's future, where other companies had merely knocked. The LCD calculator was followed within a few months by the first LCD watch, introduced by Seiko.

### Development Trajectory: Building Up vs. Building Down

Because knowledge creation is a cumulative process, achieving progressive, incremental development goals in new industry creation implies starting small and building up rather than building down from grand achievements. Past and present IBM managers recall that Ralph Gomory, IBM senior vice president and director of research, later senior vice president for science and technology, used to say "technology trickles up."[23]

Even in 2000, this idea represented a more significant assault on conventional wisdom than may first meet the eye. Since the mid-1960s, corporate strategists and business schools have been powerfully influenced by the international product cycle model of new business development. Many successful businesses, particularly in the U.S., seemed to fit the product cycle model, especially in the early Cold War years. These companies expanded around the globe by commercializing products whose development was accelerated by massive defense-related investments in science and technology that began with World War II and continued with defense and space travel. Innovations in computing, telecommunications, materials science, and aeronautics were key factors that drove this international expansion.[24] Military contracts figured prominently in the development of mainframe computers, jet aircraft, atomic energy, satellite communications, and numerous other technologically advanced goods and services. Consumer demand played a secondary role in shaping technologies and in many instances could do little more than play the role that corporations scripted for it. RCA's introduction of color television and its enforcement of a standard that all TV-set providers eventually adopted was an example of successful top-down innovation in consumer markets.

The Iridium global satellite telephone collapse provides a recent case study in why such processes may fail to work now. Motorola managers conceived the idea in 1985 of a constellation of low-earth-orbiting satellites that would provide global wireless personal communications service without regard to the remoteness of individual users. Iridium was announced in 1990, and organized as a consortium of twenty-eight partner companies and national post, telephone, and telegraph monopolies that contracted with Motorola for system development and delivery. After initial launch delays and an investment of $5 billion, the complete network of sixty-six satellites reached earth orbit in 1998. Service began in November, but customers immediately complained of poor, unreliable reception. Handsets larger than one-liter bottles cost $3,000 and worked only outdoors (when they worked at all). Calls cost up to $7 per minute. Four months later, just over ten thousand users had subscribed, providing revenues to date of $195,000. The company applied for Chapter 11 bankruptcy protection in August. Following the termination of service at midnight on March 17, 2000, Iridium faced additional expenses of $50–70 million just to de-orbit the satellites, which it was thought could not be used for any other purpose.[25] On November 20, 2000, the satellite system and Iridium's other assets were purchased out of bankruptcy by venture capitalists for $25 million. In December, the U.S. Department of Defense awarded a two-year contract for $72 million to the new concern, Iridium Satellite LLC. Service to the U.S. government began immediately.[26]

Iridium represented a grand vision and a scale of entrepreneurial risk-taking critics often regard as beyond the scope of contemporary large corporations. Its technology, strategy, and organizational innovations animated commentary among business colleagues, media observers, scientists, and scholars.[27] Perhaps with time, handsets might have been downsized, technical glitches mended, reception and reliability improved, additional functionalities identified (such as data transmission). But time—and money—ran out before Iridium could be built down to something that met its intended customers' needs.

The project failed for several other reasons that highlight the importance of building up rather than down in knowledge-driven industries. First, the project was organized to create a new technology, but without building up any corresponding knowledge of the market the consortium hoped to create and reach. The consortium incurred huge costs on untested technology before learning anything from consumers about how system or subsystem functionalities could meet their needs in practice. Second, the pace of change in wireless telecommunications technology and markets outran the Iridium concept. By the time Iridium service became available, users who valued global wireless service could obtain it readily through their earthbound cellular services, which were well developed in the geographic areas of heaviest use. Formerly atomized systems had expanded and integrated their networks, building up rather than down in response to customers' demands for access to regional, national, and international calling areas.

During the fifteen years between 1973 and 1988, the principal technical obstacles to creating CRT quality on an LCD flat panel were surmounted, one by one. Every solution along the way found application in new commercial products that would not have been possible to manufacture using CRTs. CRT replacement remained the grand objective. But companies found that progress in FPDs unlocked new functionalities, such as portability and low power consumption, that CRTs had never been envisioned to fulfill. Many of the new products sold well enough to bring profits to the companies that created them. Some products succeeded technically but failed in markets the first time out. No company failed as a consequence of bringing a flat panel device to market. In every instance, companies increased their knowledge of technologies and their market potential. By 1988, at least three companies stood on the brink of mass producing FPDs that were sufficiently advanced to bring the industry new public notice as a rootstock of future high-technology products and economic growth. Yet the application that brought advanced FPDs their first explosive growth and recognition as a real (rather than blue-sky) product—notebook computer screens—was itself a stepping stone in the path toward CRT replacement, rather than a CRT replacement technology in itself.

As Sharp's engineers worked long hours during the summer of 1973 to exorcise the "spooky calculator," another relatively small Japanese company was already preparing to introduce a product that incorporated the solution. Seiko Suwa (later incorporated with two sister companies into Seiko Epson) had also licensed RCA's dynamic scattering technology to use in making watch displays, but found it inadequate on a number of grounds. In the early 1970s, Seiko's watches enjoyed a reputation as luxury products, renowned for accuracy, reliability, and cosmetic appeal. Everyone remembered the company's status as official timekeepers at the 1964 Olympics, a role traditionally taken by Swiss watchmakers. The dynamic scattering approach to

LCD displays did not provide the reliability, degree of resolution, responsiveness, and low power consumption for long battery life that Seiko engineers and marketers found consistent with their brand's reputation.

The answer for Seiko lay in an invention independently developed by James Fergason at Kent State University and by Wolfgang Helfrich and Martin Schadt, two chemists at the Swiss pharmaceutical company, Hoffman LaRoche. Hoffman LaRoche ultimately acquired all of the patents for the invention, twisted nematic liquid crystal (TN LC) (see box following). Seiko licensed the patents and became the first company to enter commercial production, after addressing a number of challenges. These included finding liquid crystal materials that would operate for long periods at room temperature, as well as a manufacturing process that would produce reliable, long-lasting displays. Seiko engineers synthesized their own liquid crystal materials in the lab in order to do their research. Despite the uniqueness of TN LC at the time, the company moved rapidly into sales of the display component to other manufacturers in order to achieve learning and scale economies.

Like most watchmakers, Seiko guaranteed its products for one year. The company introduced its first LCD watch, the 06LC, in October 1973. It had a six-digit readout and sold for about $500. One year later, none had been returned. By 1976, Seiko's market share had grown to 60 percent of the Japanese watch market and the company was probably the second-largest watch producer in the world, as measured in unit output, behind Timex.[28] TN LC grew rapidly to dominate the LCD display market. By 1975, Sharp had also changed over and was selling TN displays to other manufacturers for use in calculators and watches.[29]

### Goal Orientation: Big Wins, Blockbusters, Small Wins, Design-Ins

Although relatively small and simple, the calculator and watch displays of the 1970s laid the foundations for the sophisticated TFT displays that established FPDs as a major industry in the 1990s. Research groups at Sharp and Seiko established the technology's manufacturability and created cash flows from both internal and external display customers that sustained their efforts. These successes created demonstration effects that encouraged larger, more

## Twisted and Supertwisted Nematic
## Liquid Crystal Displays

Twisted nematic liquid crystal displays (TN LCDs) rely for their optical properties on the manipulation of electrically induced patterns in the materials'
intermolecular arrangements. Typical liquid crystal displays have as their core
a cell comprising two glass substrates that enclose a thin layer of liquid crystal
material. The inner surface of each substrate is coated with transparent electrodes and a directional surface alignment layer to which the liquid crystal
molecules conform. When the substrates are assembled with their surface
alignment layers perpendicular to each other, the nematic, or thread-like,
liquid crystal molecules adapt by arranging themselves in twist patterns in the
intervening space. Light entering the TN LC cell follows the twists before
emerging on the other side.

Harnessing this property for display purposes requires three additional elements: polarizers, a backlight, and drive electronics. Drive electronics consist
of the transparent electrodes coated on the glass substrates, together with
attached driver chips. Polarizers are assembled on either side of the cell, and the
backlight to the rear.

When light generated by a backlight passes through an initial polarizer and
then twists through a TN LC cell, it rotates 90 degrees and emerges with its
polarization reversed. Consequently, it will pass through a second polarizer set
at cross angles (that is, of opposite polarity) to the first. Without the rotation
effect introduced by the TN LC cell, the second, crossed polarizer would block
the light entirely. The display function relies on this effect. TN LC molecules
untwist in the presence of an electric field, turning a display dark at the point
of stimulation. By directing the drive electronics to selectively introduce an
electric field at subsets of individual picture elements (or pixels), a TN LC display can be made to transmit or block light in patterns that create a visual
experience. The transmission/non-transmission effect varies smoothly with
voltage as delivered by the driver chips and electrodes.

Twisted nematic liquid crystal pointed the way toward a series of improvements in these materials, as researchers found ways to increase the degree of
twist from 90 degrees to 180 degrees and even 270 degrees. The latter materials became known as "supertwisted nematic," or STN LC. Increasing the
degree of twist improves the response time of the liquid crystal molecules.[30]

conservative companies, such as IBM, Hitachi, Matsushita, NEC, and Toshiba, to push ahead with their own LCD display programs. The early research programs used market feedback to define the functionalities necessary for displays of the future and to help identify technology trajectories that held promise of meeting these needs. New products incorporated FPDs with increased screen size, viewing angle, image clarity, contrast, vividness, and video response rates as well as reduced power consumption. Companies pursued advances in crystal materials (see preceding box), drive electronics, backlights, substrates, and other materials in order to achieve these improvements. The advent of the supertwisted approach and its successors enabled the introduction of early personal digital assistants, such as Sharp's Wizard (1987). In order to meet the promise of these new liquid crystal approaches, drive electronics remained as a principal performance frontier. The marriage of thin-film transistor, active-matrix drive technologies with highly evolved liquid crystal materials ultimately created the basis for a working flat TV sold in consumer markets—at last.

The first flat televisions to appear, however, were not the giant, wall-hanging screens of destiny. Seiko and Sharp had come to see the quest for the holy grail as a series of small contests leading to small wins. Sharp researchers operated with a slogan that emphasized the importance of incremental progress rather than creating a blockbuster: "We must surpass the CRT in performance, not in size."[31] At Seiko, the idea of going for a small TV came naturally. "Our logic was that we were a watchmaker," said Yoshio Yamazaki, who spearheaded Seiko's LCD research beginning in the early 1970s. "And to a watchmaker, small is beautiful."[32]

Seiko was first. After many months of working more nights, weekends, and holidays than even their hard-charging colleagues, Shinji Morozumi and Koichi Oguchi managed to light up a tiny TFT LCD television in their lab in 1982. The company announced the technological breakthrough and demonstrated a color prototype at the Annual Meeting of the SID in 1983.[33] The next year Seiko introduced the first commercial flat screen television, featuring a 2-inch diagonal color FPD. The new television used thin-film transistor, active-matrix drive technology (see box on *Driving LCDs*). This approach originated in 1973 at the Pittsburgh laboratories of Westinghouse, where T. Peter Brody led the team that discovered the applicability of TFTs for LCDs. Brody had devised an early active-matrix electroluminescent (EL)[34] display in 1968, using ferroelectric switches to drive each pixel. The Westinghouse group started to lose steam in 1976 after a significant portion of its funding evaporated when the company's CRT business closed in the face of competition from Japanese companies.[35] Westinghouse competitors such as Sharp and Matsushita had already recognized that the CRT would eventually fade from the market and were investing to invent and learn to manufacture its replacement.

## Driving LCDs: Passive and Active Matrix Approaches

Any display's ability to reproduce action video relies on some electronic scheme to rapidly paint and refresh serial changing images. Icon-and-cursor-driven operating systems for personal computers, most Internet applications, as well as any information or graphics-rich computing environment similarly depend on displays' capabilities to respond rapidly to changing inputs. Early electronic calculator and watch displays faced limitations on content, responsiveness, and potential size because their drive electronics consisted of rows of squared-off figure eights, each made up of electrode segments with their own drivers. In such displays, the electrodes forming each "eight" can be lit up to form any of the digits zero through nine. LCD researchers recognized early that such a system of independently wired display segments would prove impractical for larger displays or any display featuring full motion. In order to display alphabet, *kanji,* as well as numeric symbols, researchers moved toward a "dot matrix" scheme that divided FPDs into arrays of pixels. The driver electronics worked by addressing individual pixels with appropriate voltages to activate the light modulation properties of the liquid crystal (as described in previous box). Contemporary LCDs use one of two methods—passive or active matrix—to address the pixels.

Passive matrix or PMLCDs incorporate dense lattices of transparent electrodes in which each row and column intersection uniquely identifies an individual pixel. In manufacturing, the column and row electrodes are coated on separate glass substrates, the substrates assembled, filled with liquid crystal, and then sealed to form the basic optical cell. In operation, driver chips connected around the cell edges activate individual pixels by sending precisely timed and calculated voltage charges along the row and column electrodes.

PM displays encountered a technical limitation in the form of a tradeoff between picture resolution and contrast caused by a phenomenon known as "The Iron Law of Multiplexing," first described in a paper by Paul Alt and Peter Pleshko in 1974.[36] As the drivers scan the display for each video frame, each entire row containing selected pixels receives a voltage charge in sequence. At the same time, each individual pixel in the row also receives an additive voltage from the column drivers if the frame calls for a "selected" state, and a smaller voltage if it is to remain "unselected." The difficulty arises because the column drivers must pulse the entire display with a set of voltages every time they address the pixels in an individual row. Consequently during each frame, selected pixels in each row receive a relatively high voltage, unselected pixels receive a relative moderate voltage, and all pixels receive a series of pulses from the column drivers, known as cross-talk, as other rows are addressed.

All of this happens very quickly. Video frames change about sixty times per second. The rows, no matter how many may be incorporated in a display, must be scanned from top to bottom every time. But as the number of rows increases, the number of column addressing pulses also must increase, degrading the voltage differences between selected and unselected pixels. This degradation causes the tradeoff. As the number of rows of pixels or scanning lines packed into a viewing area increases, resolution increases. As the voltage differences between selected and unselected pixels decrease, the range of variation between dark and light elements of the video picture decrease. Things get either fuzzy or faded. Adding lines to increase screen size runs into the same limitation.

Over the years, engineers have achieved advances in liquid crystal materials and driver chips that have pushed out the boundaries that the Iron Law imposes on PMLCDs' performance capabilities. Current PMLCDs tend to compare favorably in performance to early active matrix LCDs, at much lower costs. Nonetheless, the premium market in most applications, as well as hopes for the future of LCDs as a CRT replacement, belong to LCDs that employ active matrix drive electronics.

The dominant approach to active-matrix avoids the problem of cross-talk by associating each pixel in a display with its own transistor. Each transistor acts as a switch that can activate a selected pixel, receive the relevant voltage for a given video frame, hold the voltage while returning to switched-off mode as other rows are addressed, and switch on again to receive new instructions when the frame is refreshed. As Steve Depp and Web Howard explained, "Active-matrix displays closely resemble dynamic random-access memory (DRAM) chips. Both are complex integrated circuits that store charge in a million or so discrete locations, each one under the control of a single transistor."[37] Engineers easily adapted the active matrix approach to color reproduction by grouping pixels in triads, associating each with a filter for a primary color.

Thin-film transistors (TFTs) dominate the field for implementing the active matrix approach. The number of transistors incorporated in a typical display exceeds one million. The TFTs are fabricated on one of the two glass substrates that comprise an FPD, using a precision lithographic process to deposit semiconductor, insulator, and electrode layers. The process requires an ultraclean manufacturing environment and has much else in common with integrated circuit manufacturing. The much larger substrate size constitutes a principal difference. Cell assembly follows the same steps described in earlier boxes.[38]

We will examine the TFT manufacturing process in greater detail in a box in Chapter 5.

Seiko's small TV debuted in the market in 1984. Yet, general amazement and critical acclaim aside, small "pocket" TVs did not at first succeed with consumers. They were relatively costly (Seiko's cost $315). As one Japanese senior manager put it, "Pocket TV? Too small to look at. Too large to put in pocket."

Commercial success or not, the Seiko flat TV marked a turning point in the emergence of the high-volume FPD industry. Major advances and new product applications for LCDs had emerged relatively slowly in the years following Sharp's and Seiko's 1973 calculator and watch introductions. High-volume production of relatively large passive matrix STN LCDs did not begin at Sharp until 1986. Sharp took three years to follow Seiko with a prototype small TFT TV, releasing a color model with a 3-inch screen in 1986. The company initiated high-volume production of the 3-inch TFTs one year later. Before another year had passed, Sharp ramped up high-volume production for 4-inch TFT panels.[39]

Sharp's 4-inch ramp-up decision shaped the platform from which LCD technology would take off to launch FPDs as a distinctive industry. The effects were felt both inside and outside of the company. The process of scaling up from 2-inch to 3-inch to 4-inch TFTs created a critical mass of experience within Sharp as well as an *esprit* among a group of relatively young engineers who identified themselves with advancing LCD technology. The company enjoyed a smashing market success with a new camcorder concept, whose developers had acted on the idea of designing in a TFT viewfinder and playback screen. Combined with orders for the 4-inch panels pouring in from other companies, the demand provided an unprecedented laboratory for creating new process knowledge at large scale. "Both inside and outside, we had a good business," Hiroshi Take recalled in 1997. "For example, we had a good customer in the (Sony) Video Walkman." Finally, the market impact of products containing the 4-inch screens loosened LCD investment purse strings at a number of larger companies. At Matsushita, the global market share leader in CRTs, a senior manager commented, "For awhile, investments might have been delayed because of the lack of internal customers. Product innovation from outside—especially the Sharp Viewcam—may have helped to motivate the decision to go ahead with LCDs."

Inside Sharp, Isamu Washizuka, who had helped lead the hand-held calculator effort more than a dozen years earlier, continued to push the LCD envelope with his teams of thirty-something engineers. Take recalled meetings each evening with Washizuka, other "upper echelon" managers, and the other young team leaders. Washizuka would hear their reports, but then instigate talk of "imaginative things . . . not the things we talked about every day." Among the imaginative things that the group started batting around: the idea of demonstrating a really large TFT LCD display of nine or more inches. As Take recalled, it was around the mid-1980s, "about the time we were still struggling with the 3-inch."

But why jump so precipitously from the path toward 3- and 4-inch? Why not continue the incremental size increases? The immediate product application was not obvious. The notebook computer loomed on the horizon, but STNs seemed likely to dominate in that market. The long-term goal of universal CRT replacement remained, but seemed far off in the future. Yet success, the group believed, would bring increased resources for their continued efforts. It would help build Sharp's prestige, recruit new young engineers to the cause, and bring in new funds. Top managers agreed to plunge ahead, and the engineers accepted the challenge.[40]

**MANAGEMENT PROCESS AND THE LEAP OF FAITH**

Managing continuous development paths requires continuous management attention. This may seem obvious, but in fact, managers trying to give attention to anything face practical limitations, including constraints on time, information, and foresight. Senior managers make choices regarding the roles and cognitive orientations they adopt toward managing within these limitations. These choices can predispose companies' outcomes in the complex, uncertain interplay of events that comprise the earliest days of a new industry (see Table 3.1).

Often a negative bias emerges when senior corporate managers frame their roles narrowly as product-driven business decision-makers rather than knowledge-driven new industry creators. This approach creates a tendency to try to reduce technology management to a basket of discrete, formal, project-level oversight decisions amenable to rational assessment within existing business models. The negative bias arises in part because existing business models and product market definitions may or may not prove relevant in the new industry. Technical initiatives may fall between the boundaries of existing businesses, functional areas, and technical disciplines. Rational calculations require analysts to fix the values of many parameters that, in fact, remain unknown over the time period that encompasses a new industry's emergence. The discussion collapses into an adversarial confrontation of optimistic versus conservative financial estimates, cross-functional bickering, and interdisciplinary competition rather than a debate about technology possibilities, alternative business models, and likely paths of industry evolution. Middle and sometimes senior R&D managers inevitably end up on the defensive. The new technology rarely gets out of the lab.

Many senior corporate and business managers have found effective roles as advocates, mentors, and participants in technology development along with their R&D counterparts. These managers do not view technology management as a series of thumbs-up, thumbs-down decisions about R&D projects. They frame technology management as a continuous knowledge accumulation process. The learning process extends beyond technology per se to encompass new business models, product markets, and industries. Senior managers draw upon and manage collaboration among multiple businesses, functional expertises, and technical disciplines to assure that new product ideas do not fall between boundaries or fall victim to the competitive interests of more powerful, mature businesses or functional units. At some point in the process they stand ready to make what many managers we met called "a leap of faith" that takes a new technology from the concept stage to the brink of becoming a new business, and then beyond.

The product-driven and knowledge-driven approaches to management process each had adherents in the early days of the FPD industry.

In product-driven environments, technology development proceeded relatively autonomously for long periods of time in isolation from the day-to-day attention of corporate and business managers. Managers not directly involved in the R&D function concerned themselves with tracking "the big picture." They tended to view the picture, however, in snapshots. Senior R&D managers made progress reports to corporate and business management, the occasions timed in accordance with established heuristics. These included closing dates of formal reporting periods, requests for budget levels that exceeded discretionary amounts, or the achievement of some blockbuster technical milestone or breakthrough.

Top managers responsible for decisions to continue or discontinue projects considered themselves functional generalists. But technology and manufacturing backgrounds rarely claimed equal standing with other areas of expertise included on their generalists' palettes. The weight of many decisions fell upon financial criteria drawn from projections of existing markets. Corporate and business managers regarded R&D managers as ivory tower specialists. In the absence of any broad-based system of collegial awareness, understanding, and accountability for technology development programs, the abilities of senior and mid-level lab managers to push projects ahead often depended on their willingness to risk their careers. Many scientists and engineers working in these environments ultimately joined entrepreneurial spinoffs or startups, where potential rewards tracked more closely with the personal risks entailed. In many instances, however, the entrepreneurs found their undertakings strapped for the resources needed to carry on. Venture capitalists often followed the same narrow, current market-oriented decision models that had driven the entrepreneurs to seek independence in the first place.

In knowledge-driven environments, senior corporate and business managers regularly engaged senior laboratory managers, middle managers, and project engineers on an informal and formal basis. They tended to view businesses as large canvases which they, along with others, were painting rather than as big pictures to be occasionally visited or brought in for viewing. All senior managers regarded themselves as generalists, with an obligation to remain current in their understanding of all functional areas of the firm including science, design engineering, and manufacturing as well as the technical aspects of their companies' products sold in markets. Similarly, specialists in junior- and middle-management roles were expected to continually grow in their understanding of the company's businesses. Responsibility for making progress in new areas of technology was broad-based and collegial. Individuals at all levels contributed and often received support to implement technical as well as business suggestions. Rather than placing their careers at risk on achieving occasional milestones, individuals were encouraged to venture their ideas on an ongoing basis. The atmosphere of collaboration extended across businesses and technical disciplines as well as functional areas.

Importantly for the future of the FPD industry, the atmosphere of collaboration ultimately also extended across companies and national boundaries.

### Top Management View: A Large Canvas at IBM and Toshiba

As Washizuka roused his young engineers to the supreme effort of creating a large TFT that could stand up in a demo with a CRT, he must have been aware that managers and scientists in other companies were talking along the same lines. In fact, at about the same time, Toshiba and IBM senior managers and researchers began talking with each other about FPD technology and business possibilities. Furthermore, they were preparing to move beyond talk and into action. In Japan, the head of Toshiba's semiconductor division, Tsuyoshi Kawanishi, was in touch with his old friend Nobuo Mii, at the time R&D head and eventually managing director of IBM Japan.[41] According to an account offered by Johnstone, he had also been in touch with Haruo Tsuji, president of Sharp, to inquire about LCD business prospects.[42] At IBM headquarters in Armonk, New York, and its nearby Yorktown Heights research nerve center, discussions were taking place that ultimately grew to involve, among others, Ralph Gomory, Jim McGroddy, Steve Depp, Web Howard, as well as Pat Toole, Senior Vice President, Manufacturing, and Mike Armstrong, group executive of communications products, which included displays. A visit from Kawanishi was in the offing.[43]

In 1985, IBM operated the longest-lived FPD business in the U.S., manufacturing 17-inch monochrome plasma display panels sold mostly for trading rooms in the financial industries. The roots of the business dated back to 1967, when IBM began working with Owens Illinois on a joint research program based on research results from Donald Bitzer's lab at the University of Illinois. By the end of 1971, IBM had leveraged results from the joint program to produce a postage-stamp-sized prototype. Around that time, IBM's display business unit identified a promising application as the visual interface for a compact computerized teller system. In 1973, small monochrome PDPs went into production in Fishkill, New York. The first displays were shipped in 1974. In 1975, the company shifted focus to larger displays, and built a new manufacturing plant in Kingston, New York. The 17-inch units started coming off the line in 1983. That same year, IBM convened the first of three task forces to assess alternative FPD technologies and their likely future impact on IBM's businesses.[44]

Early findings of the first FPD task force identified color, power consumption, and portability as critical factors that would define display preferences for IBM's future customers. At the time, neither IBM's plasma technology nor the CRT (which IBM sourced in huge quantities for its terminal and monitor manufacturing business in Raleigh, North Carolina) seemed likely to ever combine these characteristics. If the early task force findings accurately portrayed the future, then IBM's display-based businesses, with millions of unit sales worth hundreds of millions of dollars, clearly stood at risk. Yet the findings were entirely speculative, even contrary to some evidence. Human factors tests had indicated that user task performance did not measurably differ between terminals that incorporated color and monochrome displays.[45] Early portable PCs (incorporating a 9-inch CRT) had not sold well.[46]

Despite the conflicting evidence, the task force findings held intuitive appeal for several wild duck managers in research and manufacturing. They arranged to seed a few color monitors into bulk monochrome CRT orders.

Inevitably, these ended up in visible positions at customer locations, on the desks of information systems decision-makers, computer *cognoscenti,* and senior managers. Color envy grew rapidly at the customer locations, and so did color orders. *QED*: color case proved. Given the apparently insurmountable obstacles to creating color PDPs, and a judgment that PDP pixels were too large for desktop or portable use, the task force recommended that IBM shut down plasma research in 1983. Manufacturing wound down until 1986, and in 1987 the equipment was sold to Larry Weber's new venture, Plasmaco. Jim Kehoe, IBM PDP manufacturing director, left IBM to join Plasmaco as CEO. It would be more than ten years after IBM shut down the plasma research program before Larry Weber's lab at Plasmaco demonstrated a color AC PDP.[47]

### Alternating Current (AC) Plasma Display Panel Technology

Bitzer's 1999 memoir[48] described his interdisciplinary Coordinated Science Laboratory (CSL) and the birth of AC PDP. Bitzer and CSL created the computer-based education program PLATO (Programmed Logic for Automatic Teaching Operations) for the U.S. military in 1960. Working with his colleague Don Slottow and graduate student, Robert Willson, Bitzer invented the AC PDP to meet PLATO's need for a display that would support rear-projected picture images simultaneously with alphanumeric information. PLATO also needed a display that would not require external memory, but would incorporate inherent memory that would not prevent any pixel from being rewritten or erased. PDPs create images by sending electric signals to a mixture of gases (generally including neon) that break down into a plasma of electrons and ions.[49] The ionized gas or plasma emits ultraviolet radiation, which stimulates phosphors coated on the inside of one of two glass substrates to emit colored light and create an image.[50] By 1967, Bitzer and his colleagues had built a 16-by-16-cell AC PDP. The drive electronics for this display fit into a suitcase-size package, which allowed the research team to demonstrate the display outside of their laboratory. This display won the 1968 Industrial Research 100 award. The group's work formed the basis for the flat plasma TVs that began to be sold commercially in the late 1990s, as well as the most lucrative patent in University of Illinois history.[51]

The demise of plasma elevated the discussion of alternative display technologies at IBM to an even more pressing level of concern. In the context of this debate, Seiko's surprise 1983 color TFT demonstration at the SID annual meeting hit like a bombshell. Web Howard brought back a first-hand report. With it, he conveyed a strong sense of foreboding about the future of IBM's display business. The SID demonstration had clearly indicated TFT LCDs' potential to meet the anticipated market need for displays that offered portability and color. Steve Depp commented in 2000, "It was like TFT LCD had been a tiny mammalian hiding under a rock somewhere, while dinosaurs still walked the earth."

Howard urged his colleagues to take TFT LCDs' newly evident possibilities seriously and move significant resources into developing the technology. He was added to the task force, where he assumed the role of TFT advocate.[52]

Yet the best course of action remained unclear, except to Depp, Howard, Peter Pleshko (of "Iron Law" fame, see box on *Driving LCDs*), and a few other wild ducks flocking around, who began to agitate for a new research and manufacturing initiative in TFT LCDs. The task force was won over with a feasibility study that indicated the idea could be achieved, albeit for a substantial investment. In fact, the report argued, the facility should go up in Raleigh, where the display business currently assembled its CRT monitors. The proposal moved up a level and reached Jim McGroddy's desk. In principle, it won him over. He commissioned another task force, this one chaired by Marc Brodsky, a leading expert on amorphous silicon solid state physics, to investigate technical aspects. Ralph Gomory reviewed the proposal and convened a group of experts to investigate manufacturing efficiencies. He, too, was won over.

Raleigh pressed for a mandate, but wanted to remain free to investigate yet another alternative FPD technology, the field emission display (FED).[53] FEDs display images in a manner similar to CRTs, bombarding phosphors coated on an anode with electrons emitted by a cathode (see box on CRTs), but across a relatively narrow gap. Instead of using a cathode ray gun mounted at the back of a bulb to emit the electrons, FEDs apply an electric current to millions of microscopic cones deposited or etched on a glass substrate. Whereas CRTs rely on heat to generate electrons, FEDs emit electrons at room temperature.[54] Although FEDs were commercialized on an extremely limited basis by the end of the 1990s, during the 1980s the technology remained a laboratory phenomenon at best.

Reaction from other quarters further blurred the picture. Senior product division managers, asked to contribute funding to the new initiative, balked. Others (antichrists!) argued that even if TFT LCD was the right technology, the idea of manufacturing the displays ought to be shelved as too expensive and strategically inappropriate. IBM's main sources of profits derived from systems integration and central processing units (CPUs), they argued, not input/output devices. Anyway, displays were commodities, and not a source of significant value-added. If the world was really going flat, IBM should outsource FPD components manufactured by others, just as IBM had always done with CRTs. This, after all, was the approach that had brought IBM success in the PC business. Despite IBM's time-honored policy of making everything it needed for its mainframes, renegades had created the PC business in the late 1970s entirely as a systems integration operation, outsourcing virtually all components. IBM make displays? The PC Division had fought that battle and won. How counterrevolutionary!

Task force number three led by the new head of display research, Barbara Grant, responded to these issues. The report concluded that for many important products of the future, displays would serve, in McGroddy's words, "as the point of integration for all of the electronics." CRTs on desktops were commodities. But flat panel displays would ultimately comprise up to 80 percent of value-added and play vital differentiation roles for all of these products in the eyes of customers. As Tsuji had been saying for years at Sharp, "For the customer, the display is our face." The report, like the two that had preceded it, was persuasive in favor of moving ahead with TFT LCDs. IBM President John Akers came on board. Michael Armstrong's area was ready to kick in funding to get the research going.

The weight of IBM's senior management, with the exception of the PC business, had fallen solidly behind the TFT manufacturing concept. Yet, somehow, implementation remained elusive. McGroddy thought the proposal would remain stalled as long as it continued to proceed within the IBM consultation and decision-making process. He questioned the wisdom of giving the mandate to the CRT-based business in Raleigh, whose managers continued to argue the merits of FEDs. "You can't raise mice in the elephant's cage," he later commented. He became convinced that if IBM were ever to succeed in commercializing TFT LCDs, the project, like the earlier initiative that established IBM in personal computers, would have to be managed "outside the system. To chase a new market, you need a new organization, because old ones will apply restrictive criteria."[55]

IBM Japan senior managers had been following the debate with interest. They, too, believed that IBM needed to make a strong move in the direction of TFT technology. They also believed that IBM Japan was uniquely positioned to take the lead in forming the new business. Established in 1937, IBM Japan played a central role in the corporation as a source of quality manufacturing for global markets, including the United States. The company enjoyed close relations within the Japanese business community, including competitors, and a public stature on a par with the most respected Japanese companies. Its twenty-six thousand employees were drawn from the cream of Japan's talent. In 1985, new university graduates ranked it among the top five companies in Japan to join. It had consistently ranked in the top fifty, the only Japanese affiliate of an American company to ever do so.[56] Yet, Hirano recalled, "Over the years, IBM always played the mentor role with IBMJ. It was time for IBMJ to have its own initiative."

TFTs seemed to present a unique opportunity for IBM Japan to step into the role of initiator. TFT LCD industry development was evolving most rapidly in Japan, and many of IBM Japan's equipment suppliers were involved. IBM Japan's potential contribution in the technology seemed easy to explain to managers in IBM's existing businesses and to fit their business's needs. The discussions underway with Toshiba showed promise of forming a joint research project that would slash the time and cost needed to develop a relatively large TFT and scalable manufacturing process.[57] McGroddy agreed. "We had been in a discussion about the high value of IBM's capital in the consumer electronics industry in Japan, and concluded that we ought to find a way to exploit that."

McGroddy, Armstrong, Mii, and Toole informally formed "The Flat Panel Steering Committee" as a kind of shadow board of directors to advocate moving into TFT research in Japan, with an eye toward manufacturing. The idea was to steer clear of problems, and if that proved impossible, the members were "senior enough to deal with them off-line."[58]

Problems abounded, particularly when discussions with Toshiba entered the record. "For some, it was like losing our religion," McGroddy recalled. "Consorting with heathens." Raleigh managers continued to field objections in support of their demand for the FPD mandate. The controversy drew the informal steering committee more closely together. The members relied on "tremendous interpersonal chemistry" and "much continuity" of effort as they worked to bring their ideas to fruition. Even arranging for Toshiba managers to visit IBM's laboratories as part of the discussions proved to be a time-consuming negotiation within IBM.

Nonetheless, by the end of five days of meetings with Toshiba in Japan during April 1986, the direction seemed clear. The two companies would enter into two years of joint research and split the cost. The research would be conducted jointly by researchers at IBM's Yorktown Heights laboratories, IBM Japan, and Toshiba. Each company would host the project for one year in its respective facilities in Japan, starting at Toshiba, where a rudimentary R&D line was to be erected as soon as possible. At the end, each company would be free to pursue its own manufacturing plans, or to walk away. On the strength of these discussions, Toshiba engineers apparently went immediately to work designing the line and ordering equipment. The contract was officially signed and work began on August 1, 1986. One month later, the line was up and running.

### Senior Managers' Roles: Advocating and Mentoring at Corning

Sunlit days in Corning, the small upstate New York community where Corning, Incorporated makes its home, can remind a visitor of Thornton Wilder's *Our Town*. On such a day in late 1986 or early 1987, Bob Yard was running a few errands on Market Street, where he ran into a colleague, Paul Miller. Miller served as a controller in Jim Kaiser's Technical Products Division and he had just come from a meeting that had surprised him. "It was the damnedest thing he'd ever seen in Corning," Yard recalled in Tokyo in fall 1996. "Five vice presidents in a meeting for a $3 million budget decision!" In fact, Miller understated the power concentration represented in the room. Those present included Roger Ackerman, group president, Specialty Materials; Peter Booth, senior VP and president, Corning, Japan; David Duke, VP and head of Corning's research laboratories; Norm Garrity, senior VP of manufacturing and engineering, Specialty Materials Group; and James G. Kaiser, senior VP and head of the Technical Products Division.[59]

The group had come together to discuss a project that Satoshi Furuyama had been putting together with the support of Kaiser, Ackerman, and Duke. Months earlier, Booth had advocated Furuyama's transfer to Corning headquarters from the company's operations in Japan. Furuyama was the designated leader of a prospective Corning business unit that would manufacture and market glass FPD substrates. Corning research in New York and the marketing organization in Japan had followed TFT LCD developments since the early 1980s through several major turning points. Around the time of Matsushita's 1986 pocket television introduction, a number of senior managers came to envision TFT substrates as a major business opportunity for Corning.

Yet the project appeared sufficiently risky and ahead of the market so that Furuyama preferred to describe his presentation as a "business preparation plan rather than an investment request." Furuyama was working on the plan as well as running Corning's ongoing substrate export business with one other full-time person and a few part-timers. He hoped that, at minimum, the meeting would gain him the resources to add one additional full-timer. Veterans of Corning's Advanced Display Products business have been known to jest that so many senior managers attended the legendary meeting because every part-timer needed to invite everyone to whom he reported.

The truth underlying this jest, however, was that the attendees represented a global network of senior management advocacy that had emerged gradually within Corning over many years of learning about the proposed business's potential. Senior management support was vital, as the industry's potential and the technological demands of participating were entirely speculative in the middle 1980s. This remained largely true even after the first high-volume, large-format TFT fabs started up in the early 1990s. Yet at that early moment preceding the FPD industry's takeoff, each of the senior managers in the room knew that to establish market leadership, Corning would need to move rapidly to establish the new business. Well ahead of the startup dates for the anticipated high-volume TFT fabs, Corning would need to commit significant manufacturing capacity.

Corning gained experience selling LCD glass in Japan over many years, beginning in the early 1970s with sales to makers of watch and calculator displays. Corning researchers made an effort to develop extremely thin sheet glass for these applications, using a product the company was selling for use as microscope slide covers for medical laboratories. Corning had developed the underlying proprietary fusion glassmaking technology as a method of fabricating extremely thin, optical-defect-free glass without the need for grinding or polishing. Early LCD technologies, however, did not require the advanced properties of fusion glass. Most of Corning's sales for these applications continued to consist of glass manufactured using more conventional methods. But in the early 1980s, Furuyama, then working in sales for Corning Japan, noted with some surprise that several major electronics groups were placing regular, gradually increasing orders for the more advanced product. Given the known state-of-the-art in LCDs and LCD applications, it was not immediately obvious why the labs were placing these orders. Clearly, something new was happening.

Furuyama followed the paper trail and visited the labs. There he discovered a number of research programs in the early stages of investigating a new LCD technology. "We called it active matrix," he recalled, "not knowing enough to call it TFT." The company purchased a multiclient study of the new technology's future, which provided significant grounds for optimism about its business potential.

Corning Japan people intensified their contacts with the TFT researchers to learn more about the technology, as well as how Corning might apply its capabilities to contribute in the development process. Through these liaisons, Corning's U.S.-based researchers concluded that they could readily build on the innate properties of fusion glass technology to help their customers accelerate progress in TFTs. Amorphous silicon deposition for thin-film transistors required glass substrates that could withstand extremely high temperatures. In addition to defect-free optics and shatter-resistance, TFT substrates also required chemical characteristics that would remain stable in the fabrication process. Fusion-formed glass offered these properties. Furthermore, the researchers were confident that they could build on these properties over the long term to meet TFT producers' ambitions to increase display size, thinness, lightness, and visual performance.

Corning's TFT proponents on both sides of the Pacific promptly received an opportunity to test their optimism. Around 1983, Corning's Specialty Materials Group was nearing a decision to close down a fusion glass facility in

Harrodsburg, Kentucky. The plant had originally been erected to manufacture automobile safety windshields. Corning had closed that business in 1971. Harrodsburg had been kept alive doing other fusion projects, most recently sunglass lenses. At a cost between $1–2 million, the plant could be refurbished and converted to produce TFT substrates, which required a different chemical composition and needed to be formed at higher melting temperatures. Corning Japan made a case to the Technical Products Division to keep the plant open, and indeed, to eventually build a new plant.

Division staff turned thumbs down. Looking around the U.S. for evidence of progress in TFTs, they encountered little activity aside from struggling startups such as T. Peter Brody's Panelvision, which he had established in 1980, two years after leaving Westinghouse. (Panelvision was two years from production.)[60] If any FPD technology seemed slated for a big future, it was PDP, not TFT, and PDPs did not require fusion glass. In some ways, LCDs looked like an also-ran. But the businesspeople in Japan had developed first-hand relationships with the companies planning TFT manufacturing there and were aware of their progress as well as market forecasts. Corning Japan insisted that the plant must remain open for reasons of future opportunity. U.S.-based staff lacked this first-hand experience and could not figure out a model from existing market data that would yield a positive financial return.

Jim Kaiser, at the time business manager for Specialty Materials,[61] listened to Corning Japan's recommendation and agreed that the Harrodsburg plant should remain open. As a consequence of prior successes in business development, Kaiser had discretionary funds at his disposal that he could use to fund new businesses within the Technical Products Division, which served a unique function within Corning as a business incubator. Among the projects vying for funds at the time, Kaiser concluded that the advanced display project offered the highest potential for a sustainable competitive advantage. According to colleagues, two dynamic factors played into his assessment. First, Corning's proprietary fusion-formed glass technology seemed uniquely matched to TFTs' apparent technology trajectory. No competitors seemed well-positioned to follow. Second, he recognized that Corning Japan's managers had built valuable relationship capital and created significant market knowledge in their painstaking cultivation of the business. Even a technologically well-matched Japanese competitor would have faced difficulties building the same network and familiarity with market needs. For a competitor from outside Japan, the barriers appeared insurmountable. Over the objections of staff, Kaiser agreed to fund the Harrodsburg investment.

The display business group in Japan began to work directly with the technical and manufacturing people at Harrodsburg to develop the business. Around this time, the Harrodsburg plant began to produce the first generation 7059F fusion-formed borosilicate glass substrate. The 7059 product code represented subsequent generations of Corning's TFT substrates until the mid-1990s, when an entirely new product finally superseded it.[62]

The year 1986 brought further grounds for optimism, when Matsushita introduced the first pocket TV using TFTs built on Corning substrates. (Seiko's 1984 small TV had used more costly quartz substrates to withstand extremely high manufacturing temperatures used for their fabrication process.) In addition, Peter Booth assumed responsibility as president of Corning Japan. Furuyama recalled that the Matsushita introduction convinced the FPD business people in Japan that the substrate business would grow very large. But they were not yet ready to propose new capacity for the second time. Booth disagreed. He engineered Furuyama's transfer to headquarters, where he was charged with making the case not only for expanding the Harrodsburg facilities but also for building Japanese manufacturing capacity as well.

In Corning, Furuyama found himself drawn immediately into the orbits of Ackerman and Duke. LCD Market forecasts by the Stanford Resources consulting firm convinced Ackerman that the new business development process was not moving fast enough. He pulled Duke in, and the two began scheduling weekly lunches with Furuyama. Two items appeared on every luncheon agenda, according to Furuyama. "David and Roger wanted first, to make sure we were getting all the support we needed to establish and grow the new business and second, to beat us and make us move faster."

The final plan projected returns that fell short of Corning's typical targets for a new business. Ackerman put it forward anyway. "In my view," Furuyama said later, "Roger made the decision to make a bet in this business even before he saw a business plan."

Corning's new glass melting and finishing facilities for TFT substrates opened in Shizuoka Prefecture in 1988. Corning's growing TFT customer base in Japan would be served from both Shizuoka and Harrodsburg. Corning had taken the first steps toward establishing Advanced Displays as an independent, global business. The friendly company from small-town, upstate New York had seized the advantage by leveraging the first-hand knowledge of trusted managers in Japan, giving them scope to build the business through cultivating individual customer relationships. Corning had invested in building a business ahead of the game. Yet no one at Corning or any other company fully appreciated how suddenly the technology stakes would escalate, nor the investment levels that would prove necessary to cross the threshold from the laboratory to high-volume production.

### Building the Research Culture: Gambling Careers vs. Gambling Ideas

By mid-1986, IBM, Sharp, and Toshiba had each given a clear top-management mandate to researchers hoping to demonstrate the TFT LCDs' potential as a large-scale visual medium. Approval of the IBM/Toshiba joint research program had by its nature actively engaged the attention of the most senior business and technology managers in both organizations. At Sharp, managers who as young engineers revolutionized the calculator industry by introducing LCDs, had reached senior positions. They viewed continued progress in the technology as vital to the company's future.

Top management support for these companies' programs contrasted with the experiences of FPD pioneers at such companies as Westinghouse in the 1970s and Lucent in the 1990s. In both instances, the absence or departure of engaged, senior management champions placed the onus of risk for pursuing LCD research on the laboratory managers and research scientists working in the programs. At Lucent, more than two dozen staff members lost their positions when the company halted its program. At Westinghouse, the T. Peter Brody team that invented the active matrix drive system found itself scrounging constantly for internal funds, outside grants, and contracts. Often these fundraising efforts involved fending off attacks by rival CRT or semiconductor research groups whose members believed Brody's group planned to degrade their technological turf by creating inferior applications.

In the end, Brody left Westinghouse and sought venture capital to found his own company.[63]

At IBM, Sharp, and Toshiba, senior managers recognized that their ongoing attention was needed not only because of the financial resources required for technological innovation, but also because success would require organizational innovation. Organizational challenges arose in part because the TFT LCD was not really a single technology, but a set of functionalities built on a combination of technologies. As a consequence, the project teams needed to include individuals from a wide range of scientific research traditions, engineering disciplines, and product market experiences. Often these differences in background translated into conflicting styles of investigation as well as divergent notions about technology possibilities, product market possibilities, and what actions might work best to achieve objectives.

The starkest divide within the IBM/Toshiba TFT research group arose in the early days of the project between team members with CRT backgrounds and team members with semiconductor backgrounds. "With forty years of experience in CRTs," Kawanishi commented, "these engineers wanted evidence before they would learn." Semiconductor engineers were familiar with working in more uncertain conditions, in competitive environments where companies had historically invested vast resources in the face of unclear customer requirements. At Sharp, electrical engineers, mechanical engineers, and chemical engineers needed to resolve similar differences in approach. Electrical and mechanical engineers, used to working with theoretical frameworks, found themselves dizzied by the chemical engineers' seemingly unending trial-and-error experimentation.[64] Team leaders found themselves presiding over hours-long debates in an effort to forge a consensus view.

IBM and Toshiba faced the additional difficulties of melding team members from two competitive organizations into a single cooperative unit and of managing cooperation among senior managers and researchers from the two parents. The project managers worked hard to erase the boundaries of specialization in addition to those of organization. Steve Depp recalled that when Yorktown Heights researchers visited the project sites in Japan, "engineers used to pair up for experiments, and take equal responsibility for both successes and failures. Of course there was some specialization. But we thought it was good to have as little opportunity for finger-pointing as possible."[65]

The development process brought many hours of tedium and discouragement. Engineers involved in some of the earliest TFT LCD technology development characterized the process as slow and incremental. Engineers solved small problems sequentially, often through trial and error, without clear-cut, long-range plans in mind. Often they could not articulate the steps they were taking to colleagues, but relied for guidance on intuition built through hands-on experience.[66] Hands-on experience could involve spending entire days looking through a microscope to find a single defective transistor among the millions coated on a substrate.[67]

The growing numbers of engineers involved in TFT activities as well as the freshness of the technological challenges meant that all of the participants were traveling over uncharted ground. Hirano recalls August 1, 1986, day one of the IBM/Toshiba joint project: "I rounded up the engineers. Most were not experienced. I thought we might start by reading a textbook." In the early days at Sharp, Take recalled, "No one had any experience, so we did not know what was impossible." Tsuji's continued support helped preserve momentum. "For me the most impressive thing was that even when we did poorly, he praised us at the working level and, amazingly, continued to give us the money for such sport."

## WHO WON THE RACE TO THE STARTING LINE?

By the summer of 1988, both Sharp and IBM/Toshiba had developed TFT LCD prototypes measuring around 14 inches diagonally, demonstrating a potential for flat video reproduction that had seemed only remotely conceivable five years earlier. The question of which company finished first and best is subject to conflicting claims. Sharp publicly announced its achievement on June 17, as is customary for Japanese companies. IBM and Toshiba did not at first announce their achievement, which was consistent with IBM company policy.[68] Toshiba later prevailed, and the companies made a joint announcement on September 21. The generally accepted view in the industry holds that Sharp finished first, but that IBM produced a larger (14.26 inches), higher-resolution prototype.[69]

Neither company paused for long to debate the question of which had arrived first at the starting line in the race to commercialize large-format TFTs. Both had arrived at a turning point that offered the sobering opportunity to place far greater resources at risk building high-volume facilities and proving a high-volume production process. Managers and engineers in all three companies knew that fashioning large prototypes individually represented R&D achievements, but in building them in quantity represented a manufacturing challenge. The manufacturing process was, as Kawanishi put it, "very expensive, very low yield, like fishing for whales in the Mediterranean Sea."

Sharp's management questioned the logic of moving very quickly to build a costly mass production facility for large TFTs. The company was still struggling with the production process for 4-inch TFT TV displays, just emerging from factories in commercial quantities in 1988.[70] Sharp had begun to mass-produce color STN LCDs for the portable computer market in 1987.[71] In 1988, before the dominance of the Windows operating system and the emergence of the Internet, STNs seemed more than adequate for notebook computer use. Senior managers in Sharp's LCD division had a vision of mass-market applications for larger TFT LCDs, but it was hazy.[72] When they began the R&D project that led to the 14-inch display, they expected that they would eventually work up to manufacturing the products by slowly scaling up the size. They never expected that they would jump from a 4-inch to an 8-inch display. Yet, once Sharp's top managers had actually seen the 14-inch TFT LCD prototype, they could not help but imagine it as the first flat television display to reach consumer markets.

The 14-inch color TFT LCD prototype developed by the IBM/Toshiba team was presented to IBM's top management in 1988 at a meeting in Japan. Web Howard made the case that the portable personal computer market created a market of sufficient size to warrant high-volume TFT production. He further contended that color STN LCDs could not substitute in IBM's high-end target customers' eyes for TFT LCD performance. He predicted that users would be willing to pay up to five times more for a TFT LCD compared to a CRT monitor because "they provided a new platform for taking work anywhere."[73]

IBM's top management team reflected on these considerations. No doubt the displays would prove expensive. The manufacturing architecture was not developed. What was clear was that each person at that meeting wanted one of those displays.

At Sharp, the young engineers on the development team debated building capabilities to jump the TFT LCD size from 4 inches to 8 inches. "We did many calculations. Results were not encouraging," Take recalled many years later. "Leadership decided to go ahead anyway. Once a decision is made, engineers have to find all possible means. An engineer is a little like a soldier—must achieve the objective, even if sometimes die in the process."

He reflected on his years prior to joining Sharp, studying electrical engineering at Kyoto University during a time when student political activism often took precedence over studies. "I am afraid I spent most of my time in the temple dojo, practicing martial arts." He would later learn how focused energy and physical conditioning—the discipline of the dojo—would come in handy during the months he worked with his colleagues to build Sharp's first large-format TFT LCD manufacturing line.

# 4

# Knowledge and Commercialization:
# Fusing Product and Process Innovation

*The most important thing cannot be money,*
*but how information is shared.*

KYE-HWAN OH[1]

*Even talk is expensive.*

TETSUO IWASAKI[2]

**F**lash forward to summer 1996. On first impression, the generous proportions of the main assembly area in AKT's manufacturing facility overwhelm
many visitors. AKT makes chemical vapor deposition (CVD) equipment.
CVD stands at the heart of TFT LCD fabrication, acting as a principle
determinant of efficient plant scale, performing perhaps the most demanding, lengthy processes[3] among more than 200 manufacturing steps. CVD
equipment absorbs a higher proportion of the capital equipment budget for a
new fab than any other process.[4] As market trends and the search for production efficiencies have driven the FPD industry toward ever-larger substrate
sizes, CVD tools have also grown in size. So have the spaces needed to operate them, the tools' costs, and the pressures on CVD makers, as well as other
equipment manufacturers, to customize their equipment to meet individual
customers' needs.

The job description of a CVD tool entails coating profoundly thin films
of metals and chemicals on glass as part of the procedure that forms the million or more transistors built into a typical large TFT LCD cell. In order to
achieve this, a tool must integrate a variety of chemical, mechanical, and electronic subsystems, each of which in itself represents a precision engineering

challenge. Each tool incorporates robots to move the micron-thin, precision-formed glass substrates in and out of deposition chambers. As substrates grow larger and thinner, this feat grows more difficult because, among other things, the glass acquires a tendency to sag when supported around the edges. Within the chambers, a tool must uniformly deposit any of several layers of materials. The sag problem gets worse as each layer adds its almost immeasurably minute burden. AKT's CVD tools produce thin film coatings through plasma-enhanced gas-phase chemical reactions, for which the equipment must maintain glass and atmosphere at precisely calibrated temperatures. All of this must take place without generating or attracting so much as a single particle of microdust to detract from the final displays' visual perfection.

In mid-1996, the AKT final assembly area accommodated several tools in a clean, white space the size of a small airplane hangar. More than a dozen scientists, engineers, and technicians wearing long, white laboratory coats conferred around, above, and within the footprint of one of the units, mostly in animated Japanese. Some seemed engaged in hands-on inspections, adjustments, assembly, or disassembly. The tool would need to be disassembled before shipping it to the customer, and then reassembled at the customer's manufacturing site. Many members of the group would participate in both operations. There was plenty to talk about, and little time to do the talking.

Nonetheless, as three visitors, including a senior manager, entered, the room fell silent. Three members separated from the group and hurried toward the visitors, waving their hands politely.

"Sorry. No visitors, please," they called across the room. "Sorry."

The surprised hosts had reason to demur. The visitors had interrupted a technical work session that included AKT people and customer representatives. Given that AKT had supplied more than 80 percent of new CVD tools in TFT fabs in the preceding two years, a casual observer might have assumed that clients regarded the technology as generic and open. In fact the pace of industry change in the early days of large-format TFT commercialization assured the inverse.

Every customer wanted to incorporate variations that its managers regarded either as technological advances or differentiating factors that would provide a competitive edge. Some variations, particularly those that enabled larger substrate sizes, challenged the entire industry to adapt. But deploying the most advanced AKT CVD technology would not, in itself, establish competitive advantage or even parity for an FPD producer. Companies varied widely in their capabilities to integrate and optimize the many different pieces of equipment required for new fabrication lines. Even the most experienced companies at times encountered costly delays achieving acceptable yields from new tool sets.

Broad market penetration for TFT LCD applications—such as notebook computers, flat computer monitors, and TVs—demanded continuous technological advances and product innovation, at the same time as it called for prices to decline significantly. This required continuous, interdependent advances in basic research, product development, fabrication equipment, and manufacturing process innovation. Equipment and materials suppliers like AKT and Corning arrived early at the crossroads of the forces driving to establish the industry. Their decisions needed to anticipate by years the direction of technology, user preferences, even demand for the final goods that would incorporate TFTs.

The units on the shop floor represented the third official generation of large-format color TFT LCD manufacturing equipment in less than five years. Companies had differed all along in their perspectives on the functions and attributes of the final products that would incorporate FPDs. As a consequence, each generation had encompassed a variety of substrate sizes. The initial first-generation, high-volume, large-format lines had started up in 1990 and 1991. NEC had chosen to produce two 9.4-inch TFTs from one 300 by 350 mm substrate (August 1990). Toshiba had preferred 9.5-inch panels. DTI had chosen to use a 300 by 400 mm substrate, the smallest that would allow two of either 9.5-inch or IBM's preferred 10.4-inch panels (May 1991). Sharp set up a fab to produce four 8.4-inch FPDs from one 320 by 400 mm substrate (fall 1991). As AKT prepared to introduce its first product, designed for use in second-generation fabs, prospective customers' preferences remained similarly diverse.

Before deciding on specs for its third-generation equipment, senior AKT managers had canvassed customers for months. They had also helped to broker broad industry standards discussions that included FPD producers, materials suppliers, other equipment makers, and notebook computer makers. Industry consensus seemed poised to embrace a 550 by 650 mm substrate. Some companies regarded this size as optimal for fabricating six 12.1-inch notebook computer screens per substrate. AKT had acted on the apparent consensus to build its Gen 3 equipment. DTI had already achieved commercial yields on its line incorporating the Gen 3 tool set at Himeji, the first producer to do so. Sharp's new Gen 3 line would, hopefully, reach target yields any day. But, as in the past, customers were already pushing and prodding to thin down the walls of a deposition chamber here . . . alter the pivot angle of a robot arm there . . . in general, *anything* to get more panels, bigger panels, thinner panels, higher resolution panels from a single substrate fed through the new gear. Gen 3 development costs were not nearly amortized, and already the talk was of "Generation 3.x," or "Generation 3-plus."

"What size substrate will that machine handle?" the visitors asked their host as he redirected their attention to a less sensitive sector of the facility. "That's something I can't talk about," he said. "It's Generation 3. All Generation 3."

## UNCERTAINTY, INTERDEPENDENCE, AND DISPERSED CAPABILITIES

The scene just described took place in Santa Clara, California. But it could easily have taken place in Japan. Most of the players were Japanese. But AKT is a U.S. company.[5] Applied Materials, the U.S. semiconductor manufacturing equipment maker, established AKT's predecessor as Applied Display Technology (ADT) in 1991. For about five years from 1993 to early 1998, AKT operated as a strategic alliance (called AKT, for Applied Komatsu Technology) with Komatsu, the Japanese heavy equipment maker. But AKT manufactured in the U.S., using globally sourced components. The company maintains principal R&D and engineering resources in Santa Clara, California, funds basic research in outside institutions such as universities, and also relies on the specialized R&D and basic research capabilities of its global supply network. As the high-volume TFT LCD industry emerged, AKT established headquarters and the seat of its president, Tetsuo Iwasaki, in Kobe, Japan.[6]

In the world of global corporations, AKT's dispersed configuration of headquarters, manufacturing, critical R&D capabilities, and ownership appeared avant garde, particularly for a U.S. company. In FPDs, it emerged early as a normal response to the knowledge-driven nature of the new global industry. The pace and trajectory of industry evolution remained unclear in the late 1980s except to a few visionaries. But the competitive consequences of failing to move forward were growing evident to the companies that occupied the product development forefront. This forefront was clearly approaching its apex in Japan, with critical supporting knowledge streams and capabilities in the United States and to a lesser extent, Europe. Emerging final goods markets linked all of these localities. The next steps in managing new industry creation required companies to establish internal organizations as well as interorganizational networks to leverage these globally dispersed capabilities and potential market opportunities to bring about high-volume production.

In this and the next two chapters, we describe the TFT LCD-fired FPD industry take-off, in which manufacturing process development took center stage, but product innovation continued at a rapid, unrelenting pace. Building the knowledge prerequisites to participate required companies to globally manage three critical attributes of new industry emergence: technology uncertainty, uncertain market prospects, and intra-industry interdependence. Uncertainty clouds the future of any emerging industry, but in knowledge-driven competition its influence is pervasive and slow to subside as the future unfolds. Knowledge creation by nature entails discovering the unknown. The possibility and timing of technological discoveries remain uncertain, as do the shapes of emergent markets and user preferences. In TFT LCDs, the speediest paths to new knowledge that vitalized the industry traced across a global network of interdependent producers, alliance partners, suppliers, customers, and competitors.

### Technology Uncertainty

By demonstrating 14-inch TFT LCDs in 1988, IBM, Sharp, and Toshiba made a leap that, in the eyes of many, established the technology's potential to substitute for CRTs on the basis of performance, convenience, added functionality, and possibly, with time, cost. Despite this achievement, however, the technology's development trajectory remained uncertain. In order to realize the technology's full potential, companies needed to improve viewing angle, video responsiveness, brightness, and resolution, as well as continue research to enable still larger screen sizes. Other FPD technologies remained in the wings—particularly PDP—that might yet pull ahead as candidates for the mass market "TV-on-the-wall." Early TFT LCD pictures offered a crispness compared to CRTs that one manager characterized as "more like a painting than a photograph."[7] Consumers might demand more CRT-like video images before they would favor any FPD technology to replace their home televisions.

In addition to the need for continued product innovation, companies faced the need to invent manufacturing processes to produce large-format TFT LCDs at high volume. The pilot lines that existed at DTI and Sharp had served well in R&D for low-volume production and prototyping. No one knew for certain whether the process technologies embedded in these lines could be scaled up to produce high yields at high volume. Many executives and engineers remained skeptical. But all knew that only high-volume production would deliver the learning and scale economies that would bring costs down. Unless the companies could drive costs down to some multiple of CRT production costs (no one knew what multiple), deep market penetration would prove impossible. In reality, creating a high-yield, high-volume production process proved even more difficult than proponents expected.

### Uncertain Market Prospects

Although all players knew that driving yields up and costs down would prove critical to TFT LCD technology's future, the product application that would pull volume quantities of TFTs through the market did not strike all observers as obvious. Planners at both Sharp and Toshiba regarded larger TFT LCD TVs as the logical next commercial step. But 3- to 4-inch pocket televisions had never made more than a minor hit with consumers. It was not clear what the market would make of slightly larger TFT TVs. IBM did not regard itself as a consumer electronics company. CRT replacement held an important place on its product agenda, but for PC monitors on office desktops. Yet in the immediate-term U.S. market, the prospective cost for the additional functionality of a flat TFT LCD monitor over a color CRT seemed prohibitive. Except for the usual flock of wild ducks at IBM, the market potential for notebook computers remained unclear beyond a narrow target group of heavy travelling professionals.[8] In the close quarters of Japan's offices and homes, the notebook held special appeal as a space-saving device. But Sharp's managers believed that STN LCDs would suffice as notebook screens.[9]

### Managing Intra-Industry Interdependence

By 1988, IBM, Sharp, and Toshiba, all bitter competitors, had each created a technology foundation to begin independently manufacturing large-format TFT LCDs. In the process, IBM and Toshiba had also demonstrated the viability of global R&D collaboration. Their project had created significant new relationship and knowledge capital, now jointly held between the two companies as well as potential equipment and materials suppliers. IBM senior managers such as Mii in Yamato and McGroddy in New York realized that if the company elected high-volume production, collaboration with Toshiba would leverage this capital as well as the capabilities of IBM Japan and the Yorktown Heights display group. They believed that continued partnership represented the speediest route to commercialization.

At Toshiba and Sharp, managers anticipated that the new large-format color TFT LCDs would sell at high volume only if the market was globalized at the instant of its creation. Sharp prospected for global customers, such as the U.S. computer makers Apple and Compaq. Toshiba's managers did the same, but also concluded that collaboration with IBM would help globalize the Japanese company's insular management culture.[10] Kawanishi anticipated that the U.S. would play a big role in the market for final goods in any long-term TFT LCD scenario. His experience with the semiconductor trade wars of the mid-1980s alerted him to the possibility that similar tensions could arise with the U.S. government over concentration of FPD manufacturing in Japan. He hoped that establishing TFT manufacturing as a joint venture with a U.S. company would help defuse the impact on the business of any such development.[11]

Creating a high-volume manufacturing process would also require close partnerships with equipment and materials suppliers who could contribute specialized expertise, technologies, and research muscle. Many suppliers, such as Canon, Nikon, Toray, and Anelva[12] were Japanese and enjoyed long-established relationships with the companies considering TFT manufacturing. Corning had also played an indispensable role in TFT development and was well-established in manufacturing materials for smaller TFTs. In 1989, Corning was considering establishing its glass substrate business as a distinct, global business unit with authority and accountability centered in Japan. Kawanishi continued the intercompany diplomacy that helped broker the IBM/Toshiba relationship by extending feelers to Applied Materials.

Like IBM, Applied Materials was recognized as a U.S. company that had distinguished itself by successfully competing in Japan. Applied made its first Japanese sales in 1972 and organized its affiliate, Applied Materials Japan, in 1979. In 1989, Applied made 40 percent of its sales in Japan.[13] Applied personnel were already stretched thin by the company's success in semiconductor manufacturing equipment. But the strong advocacy of Applied Materials Japan President Tetsuo Iwasaki persuaded the senior management team of the necessity to meet with this important customer to investigate the new challenge.

### STRANGE LOGIC: MAKING THE DECISION TO PROCEED

According to a respected former industry analyst in Japan, "an atmosphere of euphoria" prevailed as prospective TFT LCD manufacturers faced production investment decisions in 1989. Other companies besides Sharp, IBM, and Toshiba—most notably NEC—had pursued TFT research for large-format color TFT LCD displays for many years. The 14-inch prototypes suddenly raised the stakes. The announcements altered perceptions of what could be achieved in the short term and thereby changed assessments of the pace at which commercialization of large-format TFTs would proceed. Published estimates of industry potential mushroomed to $10 billion for 1995 and $40 billion in 2000.[14] FPD industry possibilities received more mass media attention than RCA's original 1968 LCD announcement. TFT production planning may have moved from the back burner to the fast track in a number of firms. IBM's Hirano recounted an evening drink with a colleague from another large company whose management had decided to jump on the TFT bandwagon. "Strange logic," he recalled. "My friend has spent twenty three years with (the company). First half, he was very happy, working on the LCD. Management was not interested. Second half—suddenly he is too busy—the product is commercialized!" Another large company considered upgrading the priority of its TFT program despite sub-contracting FPD activities for years because of its own engineers' disinterest.

Within the companies that had announced the large prototypes, a more sober atmosphere prevailed. Despite the screens' undeniable attractiveness, indeed, glamour, new production planning models added little clarity to the picture available at the projects' beginnings. It seemed impossible to untangle the triple conundra of achieving high-volume production efficiencies, estab-lishing an attractive cost structure, and identifying markets that could absorb the output if either of the prior goals could ever be achieved. And what if another technology eventually were to catch up and take over? As Steve Depp put it, "Whatever the business plan said, it would have had to be pretty spurious."[15]

Antichrists arose once more at IBM to insist that the idea, while nice, would simply cost far too much to implement, particularly given business dif-ficulties that IBM was experiencing at the time. On September 30, 1989, IBM warned that third-quarter profits would fall short of the most pessimistic analysts' expectations. Its mainframe division, loaded down with excess overhead, found itself in a slowing market populated by increasing numbers of competitors. In October 1989, *Business Week* observed that IBM had cut prices by 60 percent to hold its mainframe market share. Further-more, the article pointed out, IBM's PC line lacked a laptop computer, "even though that $2 billion, Japanese-dominated market is growing at 40 percent a year."[16]

The joint development project with Toshiba had ended in August 1988. As top management at both companies debated next steps, Toshiba engineers opened negotiations with equipment suppliers for an up-to-date production line, as the line that had served the R&D team so successfully in the joint project's second year remained at IBM's Yamato facility.[17]

At Sharp, the young engineers who had worked on the prototype cooled their heels while continuing to investigate basic technical problems that

would help with the manufacturing. They did many discouraging calculations. Several schools of thought contended over the decision to move ahead
with high-volume production. One faction suggested that the manufacturing
process essentially equated to large-scale integrated circuits (LSIs) fabrication
and could be ramped to high volume at relatively low cost. Another faction
objected to this analogy. Yet another consisted of those who insisted
("naively," Take later commented in jest) on going ahead in any case.[18]

In each of these companies, the TFT LCD production decision ultimately
came down to a question of whether or not top management would override
the financial analyses. At Sharp, one observer suggested, top managers determined to go ahead because they regarded TFT LCD production as imperative to "fill the empty pedestal in the product hall of fame."[19] The TFT LCD
would leapfrog the color CRT, which Sharp had been forced to outsource
despite the company's history as the first in Japan to license CRT technology
from RCA. "Working for a future,"[20] Sharp would use the TFT LCD to finish the drive for independence that started with building its first chip factory
in 1970. IBM top management responded to the confidence that respected
senior managers at IBM Japan expressed, a vision of a chance to "do something really special in the portable computer market," and perhaps most of
all, the beauty of the prototype.[21]

The debates proved time-consuming. Top management decisiveness came
with a lag. Nearly one year after the IBM/Toshiba prototype announcement,
on August 30, 1989, the two companies announced their agreement to form
the DTI manufacturing alliance. The alliance was structured as a 50–50 percent joint venture between Toshiba and IBM Japan. The partners initially
capitalized DTI at about $140 million,[22] earmarking $105 million for a high-
volume TFT LCD fabrication facility. DTI's headquarters and first fab would
be located in Himeji City, next to one of Toshiba's STN LCD fabs. DTI
officially started up on November 1, 1989. Sharp management also decided
sometime during this interim to go ahead with large-format color TFT
manufacturing. Both companies announced plans to initiate production on
Generation 1 lines in fall 1991.

Corning anticipated the movement toward large-format TFT LCD
production with its early 1988 decision to add fusion glass production facilities in Shizuoka. The momentum established with this decision, which

anticipated demand by three years, proved fateful for Corning's market position. The new Sharp and DTI facilities initiated production at 12,000–20,000 substrates per month, and the companies planned to rapidly increase output.[23] Despite very low yields, the producers needed a sufficient supply of substrates to operate at capacity. Even if most production ended up as waste, the rate of process learning to increase yields varied directly with throughput. As IBM's Bob Wisnieff said, "It takes an awful lot of glass flowing through a line, really just acting as a pipe cleaner."[24] In early 1989, the substrates business moved out from under the wing of Corning's Technical Products Division to become part of the new Advanced Display Products Division of the Information Display Group. Satoshi Furuyama headed the business. He would return to Japan in 1990 to run it as president of Corning Japan, K.K. Applied Materials' discussions with Toshiba remained unresolved. IBM and DTI had also joined the talks.

Nearly three years passed between the prototype announcements and production startup: a lull before a storm. In the four years that followed, the earliest companies to enter high-volume production ramped from yields below 10 percent to above 80 percent not once, but at least three times. The industry tore through three complete generations of production technology and at least two subgenerational variants. During the same period, there also appeared the long-awaited killer application that would position FPDs to supplant the CRT. In fall 1991, Sharp's and DTI's high-volume, large-format color TFT LCD production lines were ready to start up. The FPD industry stood on the brink of taking off.

The nature, intensity, and pace of the competition that soon broke out, however, stunned the participants and confused most outsiders. Many managers and most industry observers hoped that the high-volume industry would emerge in a relatively orderly product-driven competitive environment and would evolve in predictable phases. The pioneering companies hoped to retain a competitive advantage from their innovations in large-format color TFTs for some time, scale up sizes gradually, and fairly soon establish a stable manufacturing cost basis consistent with creating a mass market. The managers knew that over time, the competitive advantages created by their product innovations would dissolve into cost-driven competition based on manufacturing process innovation. They would be ready for this development, and at the same time, ready with new products that would preserve their companies' differentiation advantages.

In fact, the product innovations associated with increased screen sizes fused inextricably with manufacturing process innovations that raised efficiency. As the new industry kindled, it quickly came to resemble a wildfire that had blown up and jumped a canyon, creating two leading edges of advance where just moments before there had been one, consuming in minutes the resources that might have withstood it for hours in less rugged terrain.[25]

# 5

## Starting From Nothing: Technology Uncertainty and Process Innovation

In September 1989 the business press in Japan heralded optimistic production projections from the DTI partners. Beginning in April 1991, the partners expected DTI to begin a ramp-up that would quickly bring production to a rate sufficient to produce 200,000 TFT LCDs a year, or roughly 16,000 displays per month, with an increase to 1,000,000 displays per year in 1994.[1] In early October 1989, Sharp announced a less ambitious target of 3,300 to 5,500 units per month for its planned autumn 1991 production startup.[2] Several months later, NEC managers announced that it would also jump into the market in 1990. In the months prior to the 1991 production startups, the consensus view among the pioneering manufacturers held that by 1995, large TFT LCD screens would reach price levels attractive in the mass market, less than $500.[3]

### THE FIRST YIELD WAR

By early 1992 the optimism had flagged, except perhaps at NEC, which had a head start of several months and was using a slightly smaller substrate. Managers there stated that they had anticipated problems all along.[4]

On May 15, 1991, DTI announced that production had started up earlier in the month, with the first TFT LCDs scheduled for shipment within a week or two.[5] But the ramp-up did not go smoothly. DTI engineering director, Hidenori Akiyoshi, commented, "We actually started from nothing. Nobody, us included, had any experience with large-format TFT LCD production. Although a test production run had been carried out by Toshiba's laboratory, a lot of unexpected problems were waiting as we ramped up. When we started production, the overall line yield was far below 10 percent, primarily due to equipment problems."[6]

In September 1991 yields at DTI reportedly hovered around 8 percent.[7] In other words, fewer than two in ten displays coming off the new line could actually be sold. Other companies were experiencing similar frustrations. By the March 31, 1992, conclusion of Japan's 1991 fiscal year, DTI had shipped a total of 30,000 displays, or about 4,200 per month. But as Sakae Arai, senior manager of LCD marketing at Toshiba, said, finding working units to ship "was like picking through the garbage."[8] At costs running $2,000–3,000 per working display, the manufacturers were shipping money out the doors. The race was on to learn how to put a stop to it.

### Killer Particles

The problems were many, but the most serious could be traced to a single, pervasive source: ambient dust entering into the production process. The importance of excluding dust particles did not surprise the production engineers. Kawanishi at Toshiba and many others working elsewhere on TFT LCD production believed that the manufacturing process would share about 80 percent of its characteristics with semiconductor fabrication. Engineers at Sharp, IBM, Toshiba, Hitachi, and NEC all believed that their companies could leverage semiconductor competencies to get a head start in LCD production and relied on their past experiences to forecast the TFT LCD ramp-up.

The similarities appeared to justify their confidence. Both semiconductor and TFT LCD fabrication used photolithographic processes to etch transistor designs on substrates. Both required immaculate cleanroom facilities. Integrated circuit fabrication required equipment, processes, and operators that could carry out their assignments without introducing contaminants, operating within cleanrooms that could maintain an atmosphere free of particles greater than 0.18 micron. Initial specifications for TFT LCD production set a far less demanding standard of 0.5 microns. Many experts expected experienced semiconductor manufacturers to find TFT production a cakewalk by comparison.

In practice, the cake crumbled. TFT LCDs proved far less tolerant than semiconductors of particles introduced in the fabrication process. Dust particles that find their way onto a silicon substrate in semiconductor production may cause one or more of several hundred chips processed on the same wafer to fail. These can be discarded, while the remainder will still function normally. The large surface area of a TFT LCD offers far more abundant opportunities for errant particles to settle, however, and it takes only one microscopic interloper to render the entire product unusable. One particle can knock out transistors controlling one or more pixels. The human eye perceives the disabled pixels as pinhole or color variation flaws in the picture. Most users find such flaws an excruciating intrusion into their fields of vision, particularly in close-to-the-eye display functionalities such as notebook computing. TFT LCD producers quickly determined that they needed to implement zero tolerance policies when it came to shipping products that exhibited such impairments. Production process engineers needed to change their thinking to strike a different balance in their particle control regimes between "how big?" and "how many?"

## Manufacturing TFT LCDs

This box will serve as a companion to those in Chapter 3, which provided brief explanations of important technical concepts that concern how video reproduction, LCDs, and some alternative FPD technologies work. In this box, we will explain some of the principal steps in the TFT LCD fabrication process. Fabricating TFT LCDs involves more than 200 manufacturing steps that fall into three main categories: deposition, cell assembly, and module assembly, sometimes referred to as packaging. For a more detailed discussion, we recommend Itoh and Yokoyama's chapter in the volume edited by Yamazaki, Kawakami, and Hori, *Color TFT Liquid Crystal Displays.*[9] This box relies on this source and others, as noted.

The process begins with two substrates of very thin, perfectly flat, uniformly thick non-alkaline glass. Corning's original TFT LCD glass, if scaled to the size of a football field, would have varied in thickness by less than a pencil's breadth.[10] The substrates are brush-scrubbed, subjected to ultrasonic cleaning, UV cleaning, pure water cleaning, isopropyl alcohol cleaning, and solvent cleaning to dislodge and carry off any organic or inorganic particles.[11] The front substrate is coated with a color filter, a color filter protective layer, and a common, transparent electrode film of indium tin oxide (ITO).[12] Color filter deposition shares common elements with the more complex TFT formation process, which we will describe next in some detail.

*TFT Deposition.* The million or so thin-film transistors that drive the individual pixels reside on the rear substrate. TFT array fabrication incorporates seven to nine layers of thin-film materials, each separately etched out into its own lattice-like pattern using a photolithographic process. (The number of layers depends on the transistor design.) These layers include silicon semiconductor material, insulators, metal electrodes, and the other TFT elements, deposited in successive steps using chemical vapor deposition (CVD), sputtering, and vacuum evaporation techniques.

As the first step in creating the appropriate pattern out of each layer of material, the thin-film-coated substrate is spin- or roll-coated with another layer of ultraviolet-sensitive resin film called photoresist, and prebaked. In the next step, the coated substrate is exposed in much the same way as a photographic print, using tools called aligners or steppers.[13] The patterns for the various transistor elements, drawn on a glass or quartz mask, function in the same role as a photo negative. Depending on the type of photoresist used, immersion in an alkaline developing solution dissolves either the exposed or unexposed elements in the pattern.[14] The remaining, undissolved photoresist protects the thin-film pattern during the etching process, which removes that portion of the underlying material uncovered in the development step. Then the remaining photoresist is removed to reveal the desired pattern in the underlying thin-film material. These steps are repeated for each of the layers of materials required by the design. At the end of the deposition phase, both the TFT array and color filter are coated with a polyimide surface alignment layer and baked.[15] Each of these steps must take place under strict temperature control, interspersed with repeated washing to remove any residues of chemicals used in the various processes.[16]

*Cell Assembly.* In this stage, the color filter and TFT array substrates are assembled together with a gap of 6 microns, at tolerances that accommodate no more than .02 micron spacing error across the entire surface of a 10-inch display.[17] The liquid crystal material is introduced into this gap. Prior to cell assembly, the molecules of the surface alignment layers on each substrate are oriented, using a brushing technique. Depending on the glass dimensions and the number and size of cells to be assembled from each piece, the color filter and TFT substrates can be cut into two or more portions.

The TFT array substrate is then coated with an electrically conducting paste and a sealant containing spherical or needle-like spacers applied to the edges and pre-hardened. Cell-gap controlling spacers, invisible to the human eye, are sprayed onto the color filter substrate at about three per pixel. The two substrates are stitched together and the sealant is hardened completely. Vacuum injection then pulls the liquid crystal into the empty cell through a small remaining aperture on one edge. Final sealing and attachment of the light polarizing films completes the cell assembly stage.

*Module Assembly.* Module assembly consists of attaching driver chips and a backlight to the cell. Driver chips send the signals that address the display's pixels via the TFTs and are attached to the edges of the TFT array. The backlight provides the source of light to be blocked or transmitted in the display's operations. In most displays, the two polarizers are positioned at right angles with each other so that sending voltage to an individual pixel blocks light to create an image (see box on page 63).

Engineers soon found, to their consternation, that the earliest high-volume manufacturing equipment and process generated particles at a rate far greater than anyone had anticipated. The process developers had refined the methods and equipment for creating thin-film transistors—particularly large-area chemical vapor deposition—from amorphous silicon technology perfected for solar energy panels. Solar energy panel performance was indifferent to particles introduced in the manufacturing process. Not so with TFT LCDs. Improved CVD equipment emerged as one of the most important challenges among many in the struggle to improve yields.[18] But the problems could not be attributed to any one piece of equipment. Akiyoshi explained that DTI's yields suffered from electrostatic charge buildup, contaminants introduced in CVD operations and on panel carriers, glass panels chipping or cracking, inferior seals in panel assembly, and out-of-spec materials.[19] "Unless we change the current production concept," commented Kouichi Suzuki, general manager of Toshiba's LCD division, "we won't be able to cut prices" to achieve mass market penetration.[20]

The high-volume TFT LCD manufacturing companies had entered into a vicious competition with each other to improve yields. Yet industry members realized promptly that the prospects for yield improvements were not susceptible to unilateral learning by managers, process engineers, operators, or any one company. The ubiquitous nature of the problem demanded solutions that were individualized yet collaborative, globalized yet deeply localized. Once the lines started up, managers, engineers, and operators on the scene needed to work together and with experts from suppliers, potential suppliers, and in DTI's case, from the parent companies' laboratories. It was necessary to investigate and reinvestigate every step, machine, material, and individual person involved in the process in a search for "yield detractors." Equipment and materials suppliers opened engineering support offices that operated around the clock within their customers' operations. Companies scoured the world for new potential suppliers. Toshiba, for example, announced in September 1991 that it had signed technical cooperation agreements with five unnamed new equipment makers from outside Japan.[21] Furthermore, the manufacturers soon realized, the competing companies would need to cooperate with each other, sharing customers, suppliers, and information to bring the new industry into being.

### Process Knowledge I: Overcoming Internal Organizational Barriers

The manufacturers could not wait for a new production concept, which, in the words of one observer, would have meant that "most of the machines that . . . suppliers like Nikon and Anelva have worked so hard to fine-tune will probably have to be scrapped."[22] The Generation 1 fabs had cost around $150 million in plant and equipment. Within months, legend had it, mountains of broken glass from unacceptable products piled up behind the fabs of the pioneering manufacturers, who were also piling up materials costs for process creation and refinement that closely approximated their investments in capital equipment. Discussions with equipment and materials manufacturers about a next generation of high-volume process technology started even before the Generation 1 fabs went on line. Success in the second generation of production equipment, however, would depend on the companies' abilities to create and retain knowledge in waking up the first.

Increasing yield figures represented the most visible measure of knowledge accumulation. In effect, each company's fab acted as a laboratory, seeking successful outcomes to experiments that would shape the next and subsequent generations of production technology. DTI President Toru Shima described the problem. "In order to take advantage of the best materials and equipment, we need first to deal with internal barriers to high yields."[23]

Invested in a fixed stock of capital equipment, managers in each company soon acknowledged that successful yield management would depend more than anything else on the people involved. Companies could not address this dependence solely by promoting learning, as a company might do, for example, by improving operator training. Training assumes an existing body of knowledge. The process was unsettled and not working very well in any case. The scientists, engineers, and operators were identifying problems and inventing critical process refinements in real time, as they did their work. The companies faced a challenge in finding ways to enhance these individualized and team-based, knowledge-creation processes. They also needed to find ways to generalize the resulting knowledge, diffuse it within their organizations, and channel it into creating the next generation.

NEC's engineers spent months working in the fab with counterparts from equipment and materials manufacturers, drawing comparisons that might help explain why any one machine should achieve different yields from another of the same type. Methodologies were invented to study operators' movements, in hope of identifying specific behaviors that might contribute to performance differences among different individuals using the same machine. To facilitate this monitoring, operators attached bar codes to each other, each piece of glass, and each machine, and submitted one and all to computer monitoring. Draconian as these measures may have appeared at the time, the operators discovered surprising sources of particle pollution in otherwise mundane behavior. Shigehiko Satoh, engineering manager at NEC's LCD fab in Izumi, expressed his hope to a touring visitor that cleanroom operators would refrain from sitting down, as doing so would release a cloud of invisible particles sufficient to destroy thousands of dollars worth of products.[24] By late fall of 1991, as other companies struggled to push yields to 25 percent, NEC claimed industry leadership by announcing that it had achieved 50 percent. Some months later in mid-April 1992, DTI's yields had reportedly reached yields of about 40 percent, while NEC claimed yields well above 50 percent, on the way to 80 percent. Other reliable sources have suggested that DTI did not reach 50 percent yields until some time in 1993. Yet another reliable source suggested that industry-wide yields had just reached about 50 percent by 1993, with DTI hitting somewhere between 50 and 60 percent.[25]

In May 1992, IBM and Toshiba named new leadership for DTI, appointing Toru Shima as president. Shima brought with him thirty years of experience as a semiconductor engineer, including terms as president of Toshiba's Tohoku Semiconductor joint venture with Motorola, and most recently three years as general manager of Toshiba's Memory Division. For twelve years, Shima had also served as an adjunct engineering faculty member at Osaka University, teaching masters' students. Delighted with the DTI appointment, Shima recalled in 1997, he considered "managing JVs, my happiness."[26] He made plans to sustain a leadership position for DTI by moving rapidly to Generation 2 TFT LCD production technology. His immediate mission: increase yields.

Shima focused on developing people as the primary knowledge carriers within DTI from Generation 1 to new generation process technology. Consensus was growing in the industry that next generation equipment would need to fundamentally differ, and DTI expected to continue to leverage Toshiba's close, long-term relationships with suppliers to participate in this process of change. Creating knowledge to carry forward in the operator force and as direct input to the next generation development process depended on raising yields as soon as possible with the current generation. Shima initiated a program to confer the degree of "Dr. Particle" on individual operators, charging them to identify all sources of contamination, develop explanations of the causes, recommend changes, and help implement process adjustments to stamp out the problems. This program, like NEC's barcode monitoring, accessed vital production process information and knowledge that was accumulating within individuals working on the line. More important, it provided operators with responsibility and incentive to transform their experiences into knowledge that could be generalized within teams and the organization. Shima credited the Dr. Particle program with a 30-percent yield increase in key fabrication steps.[27] The program embodied a philosophy that

Shima applied consistently in managing the joint venture, encouraging "highly operational discussions" at every level.

Shima also invited DTI's equipment suppliers to establish offices within the plant, requesting 24-hour on-site human support until such time as their machines met stringent objectives for up-time and major maintenance intervals. Once suppliers met these benchmarks, he would request twenty-four-hour support based in a nearby location.

The disparate interests and objectives of the partners created disagreements within DTI that opened a forum for valuable knowledge creation opportunities. As principal customers, IBM and Toshiba presented the startup with different product specifications, even in such fundamental parameters as the final product size. DTI accommodated the differing needs of its parent/customers by alternating substrate starts on the fab line. All photo-tools (steppers) contained the masks for both companies' designs. If one partner's designs required extra steps, the differences were adjusted by providing extra starts to the other. Yield differentials between the two companies' designs were "subject to intense analysis,"[28] brought to bear not only by DTI manufacturing engineers, but also engineers at Toshiba, IBM Japan, and IBM Yorktown Heights Lab. Because all designs ran on the same equipment with the same operators and manufacturing engineers, the teams could use comparisons to diagnose causes of process difficulties. Multiple designs afforded leverage against the high costs of creating the process knowledge base, as well. As John Ritsko explained, "we were learning from two designs, but paying only half the costs."[29]

DTI could also take advantage of learning opportunities based on the different capabilities of the partners. The alliance gained the services of Toshiba and IBM "fire-fighting teams," lead by senior researchers and engineers. From Toshiba, DRAM engineers came to work on the TFT array process and STN LCD engineers to work on cell assembly. IBM's Yorktown Heights display group, particularly Bob Wisnieff, worked with DTI to deploy new TFT array testing equipment. Testing TFT substrates prior to subsequent manufacturing steps reduced waste by enabling operators to remove or repair defective arrays in advance of costly subsequent steps.

Knowledge creation depended on forging an independent identity for DTI as a cohesive entity with its own people. IBM and Toshiba were learning from DTI, gaining a knowledge base of new patents that the parents would own, and securing a supply of the most advanced TFT LCDs to incorporate in their competing notebook lines. Managers in both partner companies hoped ultimately to secure a direct profit stream from DTI for the parent companies.[30] These potential gains qualified as benefits that DTI could literally ship out the door to the owners. In order to generate these benefits, DTI needed its own processes to create a distinctive culture, dynamic learning capabilities, and a growing knowledge base so that, in the words of Kanro Sato, general manager of Toshiba's LCD Division, "one plus one alliance partners can add up to two-plus, not one-point-something."[31]

DTI and the partners addressed the issue of creating a cohesive DTI entity as the medium and place through which each might leverage the contributions of the other.[32] In addition to capital and technology contributions, IBM and Toshiba had filled many initial DTI staff positions from their own personnel, especially at the management and professional engineering levels. Shima introduced policies that proscribed discussion of IBM and Toshiba goals. He asked DTI people to focus on their own organization as an independent manufacturing company pursuing a mission to improve TFT LCD manufacturing process. DTI people playfully enforced the new policy by introducing an equivalent of a U.S. workplace institution, the "swear box." Any meeting participant who mentioned either parent company was required to contribute 1,000 yen (about $10) to a large jar kept in each conference room for this purpose.[33] IBM or Toshiba groups that visited the venture or worked with it from a distance "supported DTI, not the Toshiba or IBM people at DTI." As DTI staffed up, the proportion of its members that originated at either parent declined steadily, reinforcing individuals' identification with DTI as a distinctive enterprise.[34]

As Sharp, NEC, and DTI continued to increase yields during the summer of 1992, a new set of problems emerged. TFT LCD inventories had started to accumulate as early as February. As the pipeline backed up, IBM and Toshiba reconsidered their original agreement to devote DTI's production exclusively to the partners' needs for two years.[35] In July 1992, Toshiba, IBM, and Sharp announced TFT LCD investment cutbacks. NEC pinned its growth expectations on plans to sell at least 50 percent of its TFT LCD output to external customers. TFT LCD displays still cost about $2,000 each. The TFTs competed with lower quality, much more inexpensive STN displays for use in notebook computers.[36] The yield improvements had caught the industry without plans to dispose of the increased production. Yet progress toward lower prices and a truly high-volume industry depended on continued efforts to push the lines toward full yields at full capacity.

### Process Knowledge II: Improving Equipment and Materials

Managers from the companies that had risked first-generation production could see that their attention to people and process would soon push the existing equipment to its limits. New knowledge from efforts to raise yields fed directly into new equipment and materials designs for second-generation fabrication lines. The pioneering manufacturers were anxious to embody and codify the improvements in new equipment, manuals, and process recipes, and to find fundamentally different solutions where production bottlenecks or yield detractors persisted. Widespread agreement in the industry held that CVD represented the most significant challenge.[37]

Some TFT LCD producers, like NEC and Hitachi, moved toward strategies to develop internal equipment and materials manufacturing capabilities. NEC owned Anelva, the leading CVD equipment maker, and moved into substrate manufacture to compete with Corning. Zenzo Tajima, product manager for Hitachi's TFT LCD business, increasingly dissented from apparent industry understandings regarding the next step in substrate dimensions. The company's non-conformist stance added to the negotiating challenges facing equipment and materials makers. Hitachi, for its part, seriously considered relying more heavily on internal sources, or requesting help from *keiretsu* partners such as Kokusai (21-percent Hitachi shareholding), a CVD equipment maker.

Strong commitments to speedy cost reduction formed the basis of both companies' inclinations to turn inward. NEC executives believed that TFT LCD manufacturing costs should amount to no more than double the costs for equivalent size STN LCDs. They hoped internally produced materials and equipment would enhance process integration and innovation, leading to more rapid cost reduction.[38] NEC enjoyed a reputation in the industry for yield management and cost control. Tajima believed that that TFT LCDs should cost no more than twice as much as CRT monitors. Hitachi calculations suggested that the company could increase substrate sizes ahead of the industry, while maintaining the same per unit processing time. Assuming yields and the size mix of TFT LCDs held constant, this would increase manufacturing efficiency, lower costs, and raise output.[39]

The early DTI experience caused senior managers in Toshiba's semiconductor business to recall their 1989 request to Applied Materials for TFT CVD technology recommendations. Toshiba management had not acted on the resulting proposal, which had rested on the shelf for some time.

In response to Toshiba's original request, Applied had posted a high-level management and technology delegation to Toshiba's existing small-format TFT LCD facilities for an assessment and discussions. The delegation included Iwasaki; James C. Morgan, Applied Materials chairman and CEO; Dan Maydan, executive vice president;[40] Dennis Hunter, managing director of corporate development; and the company's chief technical officer.

Although opportunities in Applied's expanding semiconductor business already placed pressing demands on its resources, particularly people, management determined to expand operations to enter the TFT LCD business if Toshiba requested it. The new industry presented an opportunity to build on the company's existing semiconductor competencies with an eye toward that industry's future. As semiconductor wafers increased in size, Applied expected to leverage new knowledge it would develop in large surface area deposition for TFTs. Failure to accept the challenge would disappoint a valued customer and leave the learning opportunity open to a competitor.[41]

The visitors from Applied observed that the fab's existing solar energy panel-based CVD technology appeared to optimize on the wrong parameter for TFT LCDs. The process designers had emphasized substrate throughput over particle control. The incumbent technology offered advantages. It built on an existing knowledge base, having created something of a sensation when a number of companies, including Toshiba and Sharp, had incorporated it into such products as solar-powered calculators. The process operated at relatively low temperature, 200 degrees centigrade, which diminished the potential for introducing flaws into the glass substrates. But the tools operated on a large-batch system similar to an assembly line and generated particles as the substrates moved from step to step. The substrates moved through the equipment four at a time on trays. The trays became coated at the same time as the substrates, but then flaked because their coefficient of thermal expansion differed. The trays traveled on racks, which also generated contamination. The system required frequent down time for cleaning, which slowed production for the entire fabrication process. Toshiba managers wanted a new CVD approach that would maintain or increase throughput, reduce down time, but cost no more per tool than the existing equipment.

After considering their customer's focus on throughput, Applied's engineers began to think about how they could apply semiconductor-based approaches to design a new, ultraclean CVD process and manufacturing tool. By massively reducing particles, such a tool could operate at higher yields and sustain the same usable output for reduced throughput. The engineers further determined that if they could increase deposition rates by a factor of ten while implementing their initial ideas, the new equipment might also sustain the same rates of throughput. This seemed likely to present a major challenge, but the engineers were confident they could do it. Equipment cost did not appear likely to present an issue, as the existing tools were relatively expensive. But Applied senior managers considered helping customers control costs central to its development philosophy and believed the company could design equipment that would make a major contribution in this area.

In 1989, Applied offered a proposal based on this new thinking. Toshiba responded at first that that the equipment appeared more technically sophisticated than necessary. Iwasaki was disappointed, but insisted on the value-added of Applied's recommendations. Having reached this friendly impasse, Toshiba recommended Anelva's CVD equipment for DTI's first line, while Applied and a "somewhat relieved" Iwasaki returned to concentrate on the company's overwhelming semiconductor commitments.

When the proposal revived, Iwasaki determined to broaden the customer base. He set out to persuade Sharp to join with Toshiba, DTI, and Applied in the joint development project. "At the time, Sharp was 'Gulliver' to the industry, with 60 percent market share."[42] Having persuaded the companies regarded as ranked first and second, Applied prepared to move forward. In 1991 the company formed a wholly-owned affiliate, Applied Display Technology (ADT), to manufacture TFT array production equipment. ADT located its global headquarters in the Kansai region near its key customers, DTI and Sharp. Dan Maydan was appointed chairman of ADT. Iwasaki, chairman of Applied Materials Japan, took office as president of the worldwide organization. Staffing needs played a critical role in the decision to establish manufacturing and R&D facilities near Applied Materials head-quarters in Santa Clara, California. Applied Materials and ADT both relied on recruiting and retaining a continuing stream of chemists and physicists trained at the University of California, Berkeley and Stanford University to staff their R&D projects.

Throughout 1991 and 1992, ADT and its development partners ex-changed a steady stream of visiting engineers, who worked together in the lab and on the manufacturing floor to understand the processing requirements for TFT LCDs. While ADT invested 80 percent of the funding necessary for the development project, Toshiba also committed funds and, along with DTI and Sharp, a wide range of other resources to ensure that the CVD R&D effort would succeed. Most critical among these resources, in the words of Iwasaki, the customers "intermediated between the overall manufacturing process and the CVD R&D effort"[43] by providing a continuous flow of information and first-hand interaction. The new equipment used automated cleaning and substituted a cluster approach for the in-line process. Carrier trays were eliminated, as a robot arm transferred individual substrates among vacuum deposition chambers for successive steps.

The interactions intensified in 1993 when ADT shipped prototypes to beta sites on the first generation lines at Sharp and DTI. The prototypes ran seven days per week, three shifts. These tests were necessary not only to stress-test the equipment, but to optimize the new tools in the context of the entire TFT module manufacturing process. "You need a commitment like this in order to move up the learning curve quickly," Iwasaki said. "It is critical to have data back from the customer to further the development process."

In June1993, Applied Materials announced the creation of the AKT joint venture, for which Komatsu. Ltd., purchased 50 percent of ADT.[44] The motivation for forming the alliance did not revolve around conventional joint-venture criteria of market, technology, or capital seeking. The R&D was complete, the funds invested, and Applied Materials already enjoyed a presti-gious position in Japan's semiconductor industry. The ADT venture itself was prewired into the two largest potential customers. Rather, Applied actively sought an alliance partner to address the people requirements of sustaining and growing the new business. Applied offered to ally with Komatsu after a rigorous search for a partner that shared its philosophy regarding the necessary marriage of cost effectiveness, quality, and technology advancement in the new industry. Iwasaki expected Komatsu to "have many talented young people to move quickly into this area."[45]

The industry would grow rapidly (no one at Applied guessed how quickly at the time). In order to keep pace, the new venture would need to rapidly expand its personnel to conduct site installations and testing as well as service its machines at the customers' premises. Feedback from this continuous person-to-person and person-to-equipment interaction at the operational level would play a vital role in accumulating and transmitting knowledge for new equipment generations.

In October 1993, AKT announced that its commercial CVD tool, the AKT-1600, was ready for delivery six to eight months after an order was placed. Beginning with the startup of the first Generation 2 manufacturing lines in mid-1994, all TFT LCD producers had the opportunity to benefit from the innovations incorporated in the new tool. The AKT-1600 sold for about $5 million for a four-chamber production system, which could process approximately forty substrates per hour. High-volume Generation 2 fabs needed about four of them.[46] By year's end, AKT had captured CVD market leadership by a wide margin.

The AKT-1600's introduction marked an historic moment in the new industry's evolution. The equipment spanned a productivity gap that impaired TFT LCD technology's promise as the first FPD to challenge the CRT for display market dominance. Working together to resolve the CVD problem, Applied and its development partners took the final steps to stabilize a high-yield process paradigm that many companies could—and did—emulate. The stage was set to establish a highly competitive industry structure in which many competitors would push production costs relentlessly downward to seize the mass market.

But cost leadership would prove illusory as a solitary focus for strategy. The competition to differentiate products to meet users' rapidly evolving needs had already rekindled.

# 6

## Surviving the Killer App: Market Uncertainty and Product Innovation

The AKT-1600's overnight success obscured a chaotic final design process that presaged continuing challenges to TFT equipment and materials makers' profitability. Sharp engineers preferred to optimize for 8.4-inch displays, while Toshiba product planners preferred 9.5-inch displays. AKT planners thought they had better build in a margin for further variation in preferences, so the designers drew up a machine to accommodate a 350 by 450 mm substrate. In the mean time, Toshiba product planners started to appreciate IBM's preferred dimension of 10.4 inches. IBM engineers figured out how to manufacture four 10.4-inch displays on a 360 by 465 mm substrate. Hirano, IBM's technical representative to DTI, debated with his counterpart at Toshiba about the wisdom of requesting a machine that would accommodate this dimension. Toshiba's engineers estimated that such a request would set AKT's development process back eighteen months. Hirano opened up a back channel. He called an ex-IBMer who had joined AKT. Hirano's colleague opined that AKT could achieve the necessary modification in six months. Ultimately, AKT's engineers thinned the deposition chambers' walls by 15 mm and changed the pivot point of the robot arm.[1]

Shortly after, Hitachi approached AKT for a design to process a 400 by 500 mm substrate for four 11.3-inch displays. Tajima believed that Hitachi should target larger displays to grab a lead in creating a flat desktop monitor market. Hitachi ended up with a CVD tool that could process 370 by 470 mm substrates, a modification of the DTI design. By the end of the Generation 2 life cycle, AKT had modified the 1600 to accommodate 400 by 500 mm substrates for a Generation 2.5 line that Sharp started up in July 1995. Even the largest market shareholders remained indecisive regarding the best display size to manufacture and the best substrate size from which to manufacture it.

## SIZE WARS

The uncertainty arose because large-format TFT LCD commercialization pushed companies in the FPD industry to engage the market dimension of knowledge creation as they had never done before. After two decades of incremental progress, building up from primitive demos through rudimentary watch and calculator LCD displays through tiny TVs, a technology had reached the threshold of the dream that started it all: giant, wall-hanging, flat television. The product application had arrived that would mediate the FPD's transition from niche market curiosity to the commonplace world. The notebook computer would spark the industry's takeoff. But this did not appear obvious at first.

As the first few high-volume lines came on stream in 1990 and 1991, the TFT LCD development frenzy remained fundamentally technology-driven. Even at IBM, where wild ducks had long dreamed of the portable work station, no internal product division had committed to purchase DTI's output by late 1990. Sharp's senior managers kept the vision of a large FPD for CRT replacement in the foreground of thinking for its young engineers. Yet company members could not reach consensus on how to create a mass market to drive progress toward high volume, lower prices, and larger TFT displays. TFT LCDs moved forward independently of any particular short-term product application because, as Take said, "We must always create new markets. That's a constant. We have to be ready to make what's next. We have to be in front of the way to make new products . . . that's our basic stance."[2]

Portable computers, also known as laptops, notebooks, or luggables, had existed in the market since Tandy Corporation's GRID introduced the concept in 1986 with a product that used a Sharp LCD. But when TFT production began, the notebook market was small. Most users who could choose preferred to work at their desktop computer rather than on a notebook. The idea that high-quality displays might alter users' preferences to favor notebooks did not capture most product planners' imaginations. Notebook computer users did not have choices. They were always on the road. They could live with submarining cursors on their unresponsive STN displays. They were happy just to have something with a keyboard.

Even as active matrix color FPDs grew more available, planners and commentators did not necessarily equate the term TFT LCD with "quality notebook computer screen." IBM's 1991 P75, based on Intel's 80486 chip, was announced at $18,000, and offered the option of a color TFT LCD or monochrome PDP from Plasmaco.[3] Many questioned whether the TFT LCD's extra performance attributes justified paying a higher price than for an STN. Sharp planners continued to believe that advances in STNs would sustain this technology in the forefront of the notebook market. The Nomura Research Institute analyst Norihiko Naono suggested that the price differential between STN and TFT displays should cause manufacturers to reflect on their expansion plans. "Dramatic improvements in contrast and uniformity have opened a new market for STN color displays," he commented. "Think about it. If you had the choice between a notebook PC with a $100 passive display and one with a $1,000 active display, which would you buy?"[4] Apple elected to manufacture its PowerBooks with monochrome TFT LCDs made by Hosiden.[5]

It remained to IBM to introduce the product that created a durable link in users' minds between color TFT LCDs and high-end notebook computing. The company had tried in vain to create an upmarket notebook segment since 1986, when it introduced a PC convertible, codenamed "Clamshell." The battery-powered computer offered many advantages over competitors' models for business people and educators, with one exception. The Clamshell used smaller floppy disks than desktop computers, making it painful for users to migrate. IBM lost an IRS contract for 15,000 units to a desktop-compatible Zenith Data Systems model.[6] The twenty-pound P75 floundered on weight and price. Even Nobi Mii refused to carry an IBM notebook. "On the road, I carry a telephone only," he commented. "The notebook is too heavy."[7]

On September 3, 1992, IBM concluded a one-year restructuring of its PC division by announcing the formation of IBM Personal Computer Company (PCC) as a separate operating unit, headed by Robert Corrigan. Corrigan set out to "fashion a new development process, in keeping with the velocity of this industry; to eliminate the communications gap between marketing, manufacturing, and development, so they could operate as one; and to squash the decision-making hierarchy."[8] The unit's charter freed it to source components from the most cost-competitive suppliers, whether within IBM or outside. Observers commented that PCC's free-wheeling sourcing strategy would cause it to resemble Apple or Compaq.[9] But neither Apple nor Compaq enjoyed a sibling relationship with a TFT LCD producer. In previous senior management roles, Corrigan had supported the TFT LCD initiative. He was delighted that PCC's mobile computing division would benefit from a guaranteed supply of state-of-the-art displays.

Two months later, IBM PCC inaugurated the ThinkPad line of notebook computers with the Model 700C. The 700C attracted immediate attention not only for its computing functionality, but also as a marvel of industrial design. The DTI 10.4-inch color TFT LCD, the largest, brightest ever available, transformed 700C owners into targets of their coworkers' envy. The unit also incorporated a small trackpoint embedded within the center of its full-size keyboard to perform cursor functions. The 700C's computing capabilities were built around an Intel 80486 processor and 120 Mb hard drive.

The product's combination of performance and design values attracted attention, but the price triggered shock waves. The 700C listed at $4,350. Toshiba reacted by replacing its $5,499 T4400SXC with the $3,999 T4400C, also a 486 notebook, but with a 9.5-inch display.[10] Prices on 80386-based notebooks tumbled. The 10.4-inch display offered up to 50 percent more screen space than other color TFT LCDs on the market.[11] By December 1992, IBM PCC had received orders for more than 100,000 units.[12] The size wars had opened. But so had the second yield war.

DTI had struggled to reach monthly production of 50,000 displays since March 1991. Under the best of circumstances, 25,000 would go to IBM. To make matters worse, both IBM and Toshiba had agreed during the period of oversupply earlier in 1992 to sell part of DTI's output to third parties who were competing with them in notebooks. But TFT LCDs had found an application that was expected to grow at 70 percent per year, and at the time, 10.4-inch displays appeared likely to establish themselves as a dominant design. IBM would need to buy quite a few of them from its competitors.

This proved difficult. By the end of 1992, IBM's ThinkPad success had triggered display shortages that rippled across all notebook suppliers. IBM struggled against a two-month backlog.[13] In early 1993, Microsoft introduced Windows 3.1, which displayed 256 colors. This added fuel to the color display fire, particularly for IBM-compatible computers, which used the Microsoft operating system. IBM PCC tried to translate its notebook market smash hit into FPD buying power, offering to source 10.4-inch displays from Sharp. Sharp's facilities were optimized to fabricate four 8.4-inch displays per 320 by 400 mm substrate. The engineers declared that the Gen 1 line had achieved yields of 60 percent, with monthly output of 90,000 displays. Sharp was offering these to high-volume customers for between $800 and $900.[14] If the company switched to 10.4-inch displays, throughput would fall to two units per substrate, resulting in wasted materials, reduced productivity, and increased costs.

Sharp, DTI, and NEC revived their investment plans. NEC hoped to quadruple production from 24,000 to 96,000 displays per month by the end of 1993 with a Generation 2 line. Hosiden planned a new fab to manufacture 100,000 monochrome TFT LCDs per month by August 1993. Fujitsu announced plans to jump in by March. In July 1993, DTI's parents announced that they would invest 30 million yen or $280 million to triple capacity with a Gen 2 line at Himeji. DTI slated the new line to start up in the summer of 1994. DTI expected the TFT LCD market to continue the 70 percent yearly growth that began in 1992 through 1995.[15] Consistent with this forecast, Sharp also planned two Gen 2 lines to start up in mid-1994.

In 1992, DTI, Sharp, and NEC manufactured about one million color TFT LCDs. By the end of 1994, these companies' capacity expansions, plus fabs erected by others including Hosiden and Hitachi, were expected to yield more than 4.8 million displays.[16] The consulting group Stanford Resources forecast that sales of all types of FPDs would increase from $3.7 billion in 1993 to $4.8 billion in 1995 to $6.8 billion in 1997. Computers were expected to account for $1.9 billion, $2.5 billion, and $3.5 billion of these sales, respectively.[17] The TFT LCD manufacturers continued to publicize a 1995 pricing target for 10-inch class color displays of $500.[18] They hoped that their announcement would persuade notebook suppliers to design more color TFT LCDs into their new models. Some questioned the announcement's credibility, in light of ongoing shortages and persistent high prices.

During 1994, TFT LCD shortages continued, eroding reputations. Road warriors related notebook hunting stories at the office and in airlines' business class. The machines emerged as fashion theft items. The information systems manager for a large New York financial services company waited three months for notebook computers with TFT LCDs from Toshiba. He commented that "IBM was impossible." He considered shifting his orders to NEC, which enjoyed a reputation for meeting its demand. Analysts speculated that among the competitors, IBM led in lost sales opportunities.[19]

As notebook market share leaders, IBM and Toshiba had the most to lose. In April, Kawanishi publicized the fact that both IBM and Toshiba needed additional supplies. "IBM Chairman Louis V. Gerstner, Jr. is asking every morning how may panels have been received from DTI," he commented. "Toshiba is asking the same question, as well."[20]

The shortages triggered additional capacity expansion announcements. IBM and Toshiba announced in July that DTI would invest $400 million in a new fab, replacing 10,000 square meters of mainframe and IC manufacturing space at IBM's Yasu site. The new line would start up in late 1995, and at full yields double DTI's production. Toshiba's spokesperson would not reveal the dimensions of the planned display. Sharp, NEC, and Hitachi also announced large investments. Added together with several other companies' more modest plans, these totaled $2 billion.[21]

## THE SHOCK OF SUCCESS

In the midst of these announcements, Sharp and DTI engineers started to bring up new Gen 2 lines at Tenri and Himeji, respectively. Sharp started up two lines, one each in May and June. DTI started up in June. Hitachi's new line started up in December. All of these new lines incorporated the new AKT CVD system and were optimized to fabricate four 10.4-inch displays per substrate. This new configuration increased production efficiencies for the larger size displays as well as output per substrate start. The new lines also raised throughput by reducing time elapsed at multiple process steps. Yield improvements from the new CVD tools were also expected to reduce costs. The new generation lines incorporated a higher degree of automation and required fewer operators. As a consequence, the pioneering producers could carry experienced operators forward to the new lines without turning their older lines completely over to inexperienced people. The operators who started up the new lines had contributed to the innovations incorporated there. They were keen to push them rapidly to their limits. The scientists and engineers were ready for a new challenge and anxious to try out the new equipment and advanced materials they had worked with suppliers to improve.

The expected capacity expansions and yield increases led some analysts to speculate on whether the industry should brace for oversupply in 1995. As the 1994 Japanese fiscal year neared its close in March 1995 *The Nikkei Weekly* projected that producers had supplied TFT LCDs at a rate of about 550,000 units per month, for 6.6 million for the year. The Korean manufacturers Samsung and LG had also announced plans. Although equipment orders suggested these companies might already have started up, the reality was unclear. Notebook sales in 1994 appeared to have reached 7.8 million, many without color TFT LCDs. During 1995, TFT LCD production was expected to increase to a rate of 890,000 per month or 10.68 million displays for the year. Assuming 30-percent growth, 1995 notebook computer sales would reach about 10.1 million. If every new notebook incorporated a TFT LCD, a 500,000 unit surplus would result.[22] In the early 1990s, the most optimistic forecasters had predicted that 1995 TFT LCD demand could reach 15 million. Companies had made their investment plans based on more conservative projections, but nonetheless aggressively.[23] Now the new capacity was coming on line. The stage was set for what Naono later described as "fierce fluctuation."[24]

The year 1995 began ominously. On January 17, the most powerful earthquake in sixty years struck the Kansai region, where most TFT LCD manufacturers located their plants. Nomura Research Institute's Hideki Wakabayashi reported that Compaq might face $400 million in lost notebook sales due to TFT LCD production disruptions.[25] The *New York Times* speculated that IBM would bear the worst blow, because it depended on DTI for displays. Jeff Cross, manager of product and technology issues for IBM, stated that that he could not verify cleanroom status for the DTI plant, but he denied trade press and competing manufacturers' claims that ThinkPad notebook production would fall three months behind.[26]

TFT LCD manufacturers exhibited little evidence of a slowdown after the quake. New product announcements evinced a trend toward larger displays that resembled IBM's favored 10.4-inch. In February, Sharp started to sample 10.4-inch TFT LCDs so customers could design them into new notebooks. Sharp's engineers had figured out how to fit 10.4-inch displays into notebooks designed for 8.4 inches, by changing drive electronics attachments to increase the viewable area. NEC engineers had learned how to produce four 10.2-inch displays on a substrate designed for four 9.4-inch displays. The company promoted these as a lower-cost substitute for 10.4-inch. Samsung announced that its plant in Kiheung was ramping up to 10,000 10.4-inch displays per month and would increase to 20,000 later in 1995.[27] LG also started production. IBM tested prospects for the desktop, introducing a 16.1-inch TFT LCD monitor for 1.5 million yen (about $15,000).[28]

In April, Stanford Resources' David Mentley anticipated that the 10-million-plus TFTs that manufacturers asserted they would make in 1995 would far exceed demand, which he forecast at about four million.[29] But prices held steady through June, when 10.4-inch displays sold for $952.50, according to Stanford Resources. In fact, the rate of growth in notebook sales slowed in 1995, but not to the extent the most pessimistic forecasters had foreseen. Unit sales grew 27% from 7 million to 8.9 million units, somewhat less than many analysts assumed. Although demand factors played a role, the slowdown was exacerbated early in 1995 by Intel's design problems with its mobile 120 MHz Pentium processor. Later, notebook suppliers faced additional delays when Intel experienced thermal problems with its 133 MHz chip.[30] By September, prices had dropped to $811.50, and in December, to less than $600.[31] Fall spot market prices from Korean suppliers were reported as low as $475–$515, particularly in Taiwan, home to many notebook assembly contractors. Samsung was reported as "most aggressive."[32]

Prices were reported near $500 in the closing days of 1995 by technical marketing consultant William C. O'Mara, who declared in February 1996 that TFT LCD production volumes had "reached liftoff."[33] Street-level discussion at trade events in early 1996 reflected a widespread impression that the industry was imploding due to the rapid price declines. Some industry officials likewise expressed concern about a possible crisis.[34] Others debunked any such ideas. "This whole biz of glut, I think, is over-dramatized," commented

Omid Milani, senior product marketing manager for NEC's LCD business unit.[35] Others spoke darkly of "market manipulation."[36] Still others called for a reality check. Rick Knox, manager of portable technology for Compaq, pointed out that a 5-percent substitution of TFTs for CRTs in the display market would wipe out industry capacity.[37] Naono, the iconoclast, had first warned in 1990 of oversupply in 1995. "Only when TFT LCDs are competitive with CRTs can earlier projections hold true," he said late in October 1996. "So far, it's not happening."[38]

In any case, the fabled $500 price point had been breached, and confusion reigned about the meaning of this achievement. Notebook suppliers like Compaq knew for certain the prices they were paying, and TFT LCD merchants like Hosiden knew their costs. Merchant/integrators, who assembled notebooks, manufactured TFT LCDs, and bought and sold them in the market, knew both prices and costs. Equipment and materials makers played vital roles in the knowledge accumulation and diffusion process that was driving yield and capacity expansions. They also understood the underlying manufacturing efficiencies. But they enjoyed the least control over their own destinies.

### Talkin' 'Bout *My* G-G-G-Generation[39]

Sharp and DTI opened discussions with AKT regarding Generation 3 equipment even as their Gen 2 lines were being installed. Sharp's engineers professed an even greater interest in pushing the 1600 series a bit further. They hoped to steal a march on the other producers by rapidly finding a way to fabricate four 11.3-inch displays on a substrate. Corning could provide a 400 by 500 mm substrate. If AKT could find a way to thin the walls of the deposition chamber and enlarge it yet a bit more, the new dimensions would fit. The Generation 3 equipment would require 550 by 650 mm substrates, to enable fabrication of six-up 12.1-inch displays. But if Sharp could capture notebook design-ins and consumer preferences with premium-priced 11.3-inch displays before the 12.1s achieved wide availability, it might enjoy a larger window of profitability from Gen 2 as 10.4-inch prices declined and Gen 3 yields slouched unprofitably through the ramp-up phase. The engineers expected to readily bring up yields for the "Generation 2.5" lines, because the equipment would represent only an incremental change from Gen 2. Product planners reasoned that if Sharp reached the market first with larger displays to coincide with the release of the new Windows 95 operating system, their 11.3-inch size would enjoy a long life.

This final twist in the Generation 2 saga underscored the need for equipment and materials suppliers to undertake broad consultation with merchant/integrators and other notebook suppliers before moving on to new display dimensions. Iwasaki undertook a comprehensive diplomatic round. He learned that the PC companies would ask for the largest displays possible.[40] This axiom was not obvious. NEC and Hitachi, for example, both ranked price above size in their planning criteria through the mid-1990s. In the early 1990s, Sharp engineers focussed on minimizing power consumption to extend battery life. Sharp stuck with 8.4-inch displays, in part, because the smaller backlight consumed dramatically less power than that of a 10.4-inch display. Weight had always been an issue. IBM was perhaps the first notebook supplier to explore product attribute preferences with focus groups of users. Steve Depp articulated the findings at an industry forum held by the University of Michigan College of Engineering in November 1994. "You ask people what they like in our ThinkPad notebook, and one thing they like is the screen . . . what you carry around for your mobile computer is basically the display."[41] Users focused on brightness, image quality, and . . . size.

Iwasaki's broad consultation did not produce a consensus, but rather a moving target. Displays would continue to grow larger to some as-yet-undetermined limit. Portability had to matter at some point. One theory held that a user's ability to open a notebook display in airlines' coach class seating would set the ultimate limit. Many questions remained. How large, how quickly? What attributes and what values of those attributes should mark generational transitions? And then, what about the question of survival? How do you spread development costs for equipment that an industry regards as generic technology among companies that in reality demand custom tools?

Once again, AKT's marketing and R&D people were forced to form their own assumptions about substrate dimensions the next equipment generation would need to accommodate. The 12.1-inch display appeared especially promising because it offered a viewable area equivalent to that of a 14-inch diagonal CRT. The company determined to configure AKT's next generation

3500 for an average substrate size of 550 by 650 mm and a maximum of 570 by 720 mm.[42] The planners hoped that the flexibility to fabricate six 12.1-inch displays or four 13.3–15.1-inch displays would attract additional customers.

Almost as soon as production started on the new machines, Iwasaki received a visit from Tajima. Hitachi planners had also invested a lot of time considering what display size and price would hold most appeal for users. They had decided that Hitachi should skip Gen 3 just as it had skipped Gen 1. Hitachi would define Generation 4 as equipment that would handle 650 by 830 mm substrates, and jump directly to it. This substrate would serve to produce nine-up 12.1-inch displays, six-up 14-inch displays, or four 18.1-inch displays. Hitachi's analysis suggested that 14-inch would dominate notebook computers, while 18.1-inch, with a viewing area equivalent to a 20-inch CRT, would launch the company into the desktop monitor lead. Hitachi planners wanted to leverage its new "Super TFT LCD" technology, which offered the wide viewing angle that research suggested flat desktop monitor users would demand. Hitachi's research suggested that by 2000, desktop monitors would comprise 45 percent of the TFT LCD market and create about $6 billion in product revenues.

Unlike Hitachi's earlier proposal to push out the limits of Generation 2 equipment, the new idea risked scrapping the new generation before it even got off the ground. In an effort to increase AKT's revenues from the short TFT LCD life cycle, AKT had just expanded into dry etch and sputtering. In 1996, AKT held 10–20 percent of these markets.[43] Industry discussions of potential standards for Generation 4 had circled around substrates of 1 square meter. If AKT moved away from the direction it had established in these discussions, as well as those that concerned Generation 3, it would place the company's R&D investments, Gen 3 orders, and credibility at risk. Even talking of such a thing could be expensive. The outer limits for stretching Generation 3, AKT product planners concluded, were substrate dimensions of 600 by 720 mm, from which manufacturers could make six-up 13.1-inch XGA panels, four-up 16-inch XGA panels, or two-up 20-inch SXGA panels.[44] Hitachi turned to its *keiretsu* partner Kokusai to make its CVD equipment.

As 10.4-inch prices slid in the second half of 1995, Sharp and DTI people worked to bring up Generation 3 lines. The Sharp teams faced the added challenge of bringing up Generation 2.5. Generation 3 lines carried the automation introduced in Generation 2 to a level of pervasive robotization. The substrates were too large for an operator to handle. Full cassettes used to transport substrates between manufacturing stages weighed about 80 pounds. At DTI, automated, guided vehicles about the size of NFL linemen rolled majestically about the cleanroom floor, moving cassettes from one manufacturing process step to the next. The units hummed a new age tune to warn of their proximity, gently nudging the scarce human operators like large pets. "Sometimes," confessed the engineer assigned to guide our tour, "I would like to break the music boxes."[45]

Because fewer humans were needed to operate Generation 3 lines, Sharp and DTI management expected the new fabs to achieve high yields rapidly. This proved true for DTI, but not for Sharp. At DTI, experienced engineers from the Gen 1 and Gen 2 lines transferred from Himeji to Yasu to bring the new line up. The reduced requirements for human intervention allowed DTI to redeploy its knowledge in this way without diminishing yields on the existing lines. In fact, DTI had maintained a stable headcount since 1994.[46] At Sharp the effort to bring up two lines at once, along with a new array process to increase the displays' aperture ratio, appeared to have too thinly spread its experienced engineers and operators. The Generation 2.5 line did not prove to be the piece of cake anticipated. The situation with the Generation 3 line appeared even worse. Although outside observers' estimates suggested that both lines were started up in July 1995,[47] by May 1996, Sharp had conceded publicly that the Gen 3 line had proven itself a "major technical challenge" and that progress was slow. DTI's Gen 3 line was by then operating at full yields,[48] having started up sometime in the fourth quarter of 1995.

The process of preparing for the Generation 3 ramp-up at DTI had started long before the equipment was installed, when project teams of engineers and operators were formed to work with key equipment and materials suppliers. Once the new fab started up, DTI's real-time intranet information and reporting system known as D-TIMES (DTI Total Information Management and Execution Systems) linked DTI, IBM, Toshiba, and the key equipment and materials suppliers to each other and to individual machines on the manufacturing line. D-TIMES tracked the progress and status of every individual unit in process. Equipment status information was refreshed every thirty seconds, shining a spotlight on any problems caused by operator error. The system also tracked actual versus planned expenses and provided continuous routing optimization on a per unit basis.

D-TIMES maintained daily 7:00 A.M. and 7:00 P.M. reporting cycles. Previous day summaries provided information on yield, work-in-progress, and shipments to DTI management at 8:30 A.M. At 11:00 A.M., all data were stored on a web server, available to IBM, Toshiba, and key equipment and materials suppliers (suppliers also maintained representatives on site for the first six months of the ramp period). "This is real time data processing. We do not introduce any human error, and we don't do any correcting," Shima said. "The delay is to prepare a proper understanding of what is happening,"

"If data needed to be keyed into the system," observed Masaru Shiozaki, senior manager, planning, "it would be out of date."[49]

### The Coming of Age

In April 1996, Sharp, Fujitsu, and Samsung announced that they would phase out 10.4-inch TFT LCDs as a result of plunging prices, after the size hit a low of $300 per unit in March. Yet 12.1-inch displays were in short supply.[50] Many Gen 2 lines were switched to manufacturing two-up 12.1-inch displays. Merchants were getting spot prices of $950 to $1,450 for 12.1-inch displays and offering volume prices of $850 per unit to long-term customers. They could generate more revenues by producing two larger displays per substrate than four smaller ones.

Besides, as Hajime Sasaki, an executive vice-president at NEC explained, switching the old lines over was profitable due to very high yields and because the fabs were fully depreciated. In fact, NEC tried to get the jump on everyone in mid-1995 by switching two older lines in Kagoshima to 12.1-inch displays, even before DTI and Sharp could shake down their Generation 3 facilities.[51] In December 1995, NEC also started up the last Generation 2 line in Japan optimized for 10.4-inch displays.

The company may have lagged the market because senior managers in NEC's PC division advocated a rule of thumb based on their conviction that notebook buyers would not pay more than twice the price of an STN for a TFT LCD screen. Despite emphasizing outside sales, NEC's TFT LCD group had weighted input from its internal customer, the NEC PC business, more heavily than information from outside customers like IBM, Compaq, and Toshiba. NEC had started the TFT LCD unit as a second FPD source for its PC division.[52]

More interaction with outside customers might have provided the TFT group with evidence to mount an earlier challenge to NEC PC's assumptions about the importance of size and cost to notebook users. In 1995, with NEC's fabs running at 70 percent of capacity and the new fab under construction, TFT LCD group head Itsuro Adachi had traveled to the U.S. to visit customers. In conversations with notebook suppliers such as Compaq and IBM, he learned that users wanted larger, thinner displays. On the plane back to Japan, he determined to stop production and retool. Adachi resolved henceforth to learn from customers outside NEC as well as from his internal customer. At about the same time, NEC's PC division was beginning to learn the religion of larger-size displays from another of its suppliers, Sharp.[53]

The economies of production, however, were stacked against older lines making larger displays, due to materials wastage and the disadvantage in unit productivity for given throughput time. Inevitable price declines in 12.1-inch displays would hit the older lines' profitability harder than the new ones. Some notebook suppliers who could not get adequate 12.1-inch supplies settled for the 11.3-inch size offered by Sharp and Fujitsu. Prices for 10.4-inch stabilized and rose slightly as supplies began to dry up.

Hosiden invested heavily in Generation 2 fabs in 1993 and was hit particularly hard by the 10.4-inch price downdraft and the shift toward 12.1-inch. Because Hosiden was a small, pure merchant producer, management did not have the benefit of learning from an affiliated personal computer division. Hosiden had no international affiliates, and its relationships with customers were insufficiently intimate to share necessary market knowledge. The privately held company's roots were in mechanical components manufacturing. It entered TFT LCDs when capital requirements were relatively low and built its business selling monochrome displays to Apple. By 1995, the relationship had deteriorated because Hosiden had not been able to meet all of Apple's requests.[54] When the market shifted toward larger displays in late 1995, Shinji Morozumi,[55] deputy general manager of Hosiden's LCD unit commented that the shift to 11.3-inch and then 12.1-inch displays had occurred rather abruptly. President Kienosuke Kamada later commented, "Hosiden really did not see this coming."[56] Hosiden's TFT LCD revenues fell from $230 million in fiscal year 1994 to $76 million in 1995.[57]

Although Hosiden could not invest the $500 million necessary for Generation 3 equipment, its TFT LCD foray had created valuable knowledge capital. Senior managers decided to search for an alliance partner with international presence, deep pockets, and intimate user relationships to unlock the value in Hosiden's technology and manufacturing process.

Senior managers at Philips, the Dutch conglomerate, were searching for such an opportunity. Philips had set up a pilot line in 1991. The company then invested heavily in 1993 with two partners in a small Gen 1 Fab based on an alternative technology for active matrix pixel addressing that engineers hoped would leapfrog TFTs.[58] Unable to modify TFT LCD production equipment to suit their approach and cut off from the mainstream of process improvements in Japan, Philips' engineers found they could not raise the facility to commercial yields. "We paid a price in loneliness," commented J.C. Stuve, president of Philips FPD, in 1997.[59]

In 1996, the two companies formed a strategic alliance that combined Philips' market, managerial, and financial power with Hosiden's TFT knowledge: Hosiden and Philips Display Corp. Skeptics viewed the combination as the first hard evidence of a shakeout that would mark the industry's passage into maturity as a cost-driven, commodity sector of global electronics.

At IBM Japan, Takahiso Hashimoto, director of display technology, commented, "So if it is a commodity market? I believe it is, as few things are unique in this world. IBM could get screens anywhere. But we have a dream to create the highest possible resolution person-machine interface. The greatest challenge to technical skill is to succeed in a commodity market."[60] At the time, DTI was succeeding at a rate of about $20 million in profit per month. According to Hashimoto, the venture could make money on any display that sold for more than $350. DTI had taken the lead in market share for the most advanced displays in 1996. Steve Depp commented, "I for one am very happy that we're in this business. From time to time, our financial people agree."

"And then we ask for more investment," Bob Wisnieff added.[61]

In July 1997, TFT LCD sales growth slowed. The September 1997 *DisplaySearch Monitor* reported, "Japan's streak of nine consecutive months of greater than 45-percent growth in LCD production ended in July according to the Ministry of International Trade and Industry as growth fell to 35.5 percent from over 50 percent."[62] The slowdown resulted in part from supply-side factors. In the wake of Sharp's difficulties ramping up its Gen 3 line, rumors of Hitachi's jump to a larger substrate, uncertainty over Korean manufacturers' plans, and 1995's rapid price declines, many companies delayed their investment plans.[63] Many managers believed that the best hope for long-term profitability lay in finally moving aggressively on the CRT's traditional domain, not in home televisions, but desktop computer monitors. Few expressed willingness, however, to invest in the learning necessary to drive performance up and costs down ahead of the inevitable price declines. Many preferred to let other companies invest in proving new process technology. They gambled that the major challenges of Generation 4 would somehow be surmounted in the course of commercializing Generation 3.

Hitachi charted its own course, moving ahead while appearing to skip Gen 3. The company internalized major elements of equipment manufacture and implemented its plans to fabricate on larger substrates. In order to free its engineers and operators to work on the new project in 1996, Hitachi contracted out 11.3-inch production to Hosiden.[64]

Notebook procurement managers forecast that prices for 13.3-inch displays would fall to the low $700s by the end of 1997 and continue to decrease through the following year as new capacity started up. Producers started to discount 12.1-inch TFT LCDs to retard a shift to 13.3-inch displays.[65]

The 1997 advent of the Asian Economic Crisis temporarily froze the industry in time. Discussion of immediate, widespread implementation of Generation 4 production lines ceased. Funds to erect new fabs dried up. The years 1997 and 1998 ended with almost identical industry sales volumes, at $10.5 and $10 billion, respectively. The early phases of industry emergence had passed. A 12.1-inch TFT LCD cost about $500, roughly one-quarter of the cost of a 10.4-inch display in 1992.

But the dream that drove the FPD industry was not frozen in time. The months ahead would offer an interlude to consolidate what had been learned and gather nerve for the quest, still ahead.

At Sharp, the aging young engineers, now in charge, announced the death of the CRT. The company released the first-ever full line of flat TVs and would incorporate TFT LCDs in its entire line of television sets by 2005. Sharp TVs smaller than 30 inches would incorporate TFT LCDs. Sets from 30 to 50 inches diagonally would use Plasma-addressed liquid crystal (PALC) technology developed jointly by Sharp, Sony, and Philips under license from a U.S. company, Tektronix.[66]

### THE CRITICAL ROLE OF INDIVIDUALS AND TEAMS AS KNOWLEDGE CARRIERS

The fortunes of IBM, Sharp, Toshiba, DTI, Applied Materials, and Corning were intertwined in the commercialization of large-format color FPDs. So too were the fortunes of every company that risked TFT LCDs as a visual path to the multimedia future. The requirement for continual innovation in both product and process during the early phases of the industry's evolution created circumstances of permanent revolution that we regard as emblematic of knowledge-driven competition. Each company played a role in the interdependent knowledge-creation process that made possible continual, simultaneous product improvements while driving costs down to reach the mass market. The most visible documents of this knowledge-creation process—increasing display size and rising manufacturing yields—represented an immense substructure of meticulously tended details and discoveries, most of which were never written down.

Although the story unfolded as competition and collaboration among great companies, individuals formed its core. Companies were destined to face the yield challenge repeatedly as they added lines to expand output and bring up new generations of production equipment to more efficiently produce larger, better-performing screens. But all knowledge creation begins with individuals. As groups of individuals gained experience working closely together, they created a shared knowledge of the TFT LCD manufacturing process. Much of the shared knowledge reached beyond company and national boundaries, extending to networks of producers, suppliers, and customers. Access to these networks enabled some companies to reduce the time and materials needed to achieve commercial yields from new lines, as well as to begin the effort before other companies could do so. Chronological time expended in these endeavors related inversely to profit. The order in which companies committed to the new generation technology often did, as well. Individuals and teams moving intellectually and physically from generation to generation acted as carriers of the knowledge needed to repeat the success.

In those early days of large-format FPD commercialization, the availability and locations of knowledgeable individuals and teams often proved important in companies' decisions about where to set up new manufacturing lines. Yet companies were also able to hire, borrow, move, or network people with the knowledge needed to set up new high-volume lines. Much knowledge became embodied in equipment, automation systems, information systems, and process recipes that could be sold to manufacturers starting up TFT LCD production for the first time. Even so, companies that lacked or failed to seek access to sufficient stocks of first-hand knowledge embodied in individuals and teams fell out of sync with the economics of industry advance. These companies suffered in the short-term or eventually fell by the wayside because they could not ramp up production at sufficient speed to drive display sizes up and costs down as quickly as their competitors.

# 7

# Knowledge-Driven Competition: Frameworks for Strategic and Organizational Innovation

**H**igh-technology competition is high-speed competition. It is also global competition. When a new technology commercializes first in a particular country, conventional wisdom assumes that local companies gain a potentially insurmountable lead over companies from elsewhere. But this did not hold true in the flat panel display industry and at best represents a misleading assumption for strategy in new, knowledge-driven industries. The FPD industry emerged as a complex global network of relationships among companies and people. Each encompassed distinctive, complementary advantages and needs. Companies succeeded when their managers challenged assumptions traditionally used to formulate new industry strategies. Access to technology and market knowledge outweighed ownership and national location of manufacturing facilities as a determinant of business performance. Companies needed to reassess strategy processes that biased managers' thinking in favor of managing projects rather than people, building physical assets rather than creating knowledge assets, producing at home rather than learning abroad, and analyzing financial results rather than managing time, the scarcest resource of all.

During most of the twentieth century, a company on the brink of entering a new industry faced the moment of truth when management decided whether or not to commit funds to build a factory large enough to produce goods at minimum cost.[1] Companies commercialized innovations by establishing manufacturing in their home countries. They projected their organizations outward to the rest of the world as market opportunities arose and to seek minimum costs of capital, labor, and materials. Vertical and horizontal integration were prescribed internationalization modes to protect firm-specific knowledge from competitors and potential competitors by sharing it only within company boundaries. Similar reasoning motivated most international companies to center scientific leadership, research, and development at home.

As the high-volume FPD industry took off in the 1990s, many companies succeeded with strategies that seemed to invert this logic. Other companies tried to play by the rules and failed. When companies entering the FPD industry chose Japan over the U.S. to establish plants, they chose distance over proximity to the U.S. notebook suppliers that would become their biggest customers. Other countries besides Japan showed equal or greater promise as economic sources of materials. The companies invested before managers identified the high-volume product market opportunities that would bring the industry to critical mass. Some accepted relatively high costs of land, plant and equipment, labor, and materials in order to locate at what appeared to be the geographic center of new industry developments. Many entered into co-development, production, and marketing alliances that required them to share vital, firm-specific knowledge with powerful international competitors. These successful companies moved decisively to create knowledge stakes in a new display functionality that offered myriad prospects in future product markets. They mobilized knowledge assets from around the world while centering their businesses in Japan, where the new industry was approaching critical mass. Their technologies and manufacturing processes had reached advanced stages of development when the high-volume, mass product markets emerged.

## CONTINUITY, LEARNING, AND SPEED

Companies succeeded or failed in the FPD industry based on how their strategies addressed three mutually reinforcing dimensions of international high-technology competition: *continuity, learning,* and *speed.* Successful companies invested knowledge assets in open, global learning processes. They sustained a continuing commitment to knowledge accumulation in FPDs through many years of difficult research, funded either through high-level corporate mandates, small product wins in consumer markets, or both. They established high-volume production facilities where knowledge was accumulating most rapidly. They acted quickly to lead developments in product markets rather than to follow. They accepted the dual challenge of simultaneous innovation in products and manufacturing processes, working to drive costs relentlessly down from day one of each new product breakthrough.

### Continuity

Commercialization of large-format TFT LCDs culminated a learning process that proceeded intermittently in all but one company each in the U.S. and Europe and continuously in several Japanese companies beginning in the early 1970s. Simple FPDs found their first commercial application as displays for inexpensive calculators and digital watches. By the mid-1970s, these low-profit products lost appeal for most U.S. companies, which abandoned the market and FPD research along with it. Several large U.S. consumer electronics companies pursued R&D for giant, wall-hanging TVs. All of these programs died out when the last of the companies disbanded their TV sales divisions and sold off their CRT factories and TV brands. Their knowledge fragmented and dissipated as researchers moved on to other projects, other companies, their own start-ups, early retirement. In Japan, FPDs gradually increased in size as companies aimed at increasingly sophisticated product applications such as personal digital assistants, hand-held TVs, and video games. The knowledge represented by these incremental advances steadily enlarged the platform in Japan for new learning and technological advance.

Following the 1988 14-inch prototype announcements by Sharp, IBM, and Toshiba, the race was on to develop manufacturing equipment, materials, and processes and integrate them into fabs to make high-volume production of large-format color TFTs possible. As technology generation followed generation, much of the process knowledge that accumulated within companies needed to be carried forward personally by engineers and operators from old lines to start up the new. Retaining knowledge depended on retaining individuals who continued with their organizations to participate in the startups of successive facilities.

### Learning

The pace of knowledge accumulation had heated up around LCD technology by the mid-1980s, particularly in Japan. This did not mean, however, that Japanese companies enjoyed a monopoly on the industry's potential. Even after the large prototype demonstrations, many technology and manufacturing problems remained to be solved before the industry could achieve critical mass and commercialize advanced display products. Notebook computers—the killer application that ultimately ignited high-volume production of large-format advanced displays—also remained in the future. Numerous companies outside of Japan owned core competencies with potential to catalyze the industry's take-off. These companies' odds of learning about industry opportunities depended first on the character of their organizational capabilities in Japan and second on a corporate mentality biased toward recognizing global potential in local opportunities and legitimizing local affiliates' global aspirations. Corning, Applied Materials, and IBM pursued FPD R&D and production opportunities that their Japanese affiliates' managers identified and advocated beginning in the early 1980s. These companies leveraged global managerial, technological, and manufacturing process capabilities to create sustainable competitive advantages through Japan-based businesses, strategic alliances, and customer relationships. As the FPD industry continued to evolve in the late 1980s and early 1990s, an international community of participants in Japan played an increasingly focal role.

### Speed

Once companies undertook high-volume production of advanced FPDs, competition in the notebook computer market and the changing demands of software and Internet applications challenged them to push product development ahead at a rate unprecedented in high technology. Notebook suppliers used screen size and image quality as a primary means of differentiating their products. Measured according to substrate area, AKT's management suggested that the rate of change in FPD technology between 1990 and 2000 exceeded the rate of change in semiconductor technology from the mid-1970s to 2000 by a factor of eighteen.[2]

Another way of looking at such data (recall Table 1.2) suggests that TFT LCD makers endured *at least* five generational changes in half the time the semiconductor industry endured the same number of transitions.[3] Japanese and non-Japanese participants alike needed technology partnerships, supplier relationships, customer relationships, joint financial arrangements, and strategic alliances more generally to stay abreast of technical developments, gain the benefits of specialization in particular aspects of FPD manufacturing and development, and to share financial risk. Strategies of pure vertical integration placed companies' learning opportunities at risk and detracted from their abilities to keep pace.

## THE SHIFTING BALANCE BETWEEN TACIT AND EXPLICIT KNOWLEDGE

As large-format, color FPDs commercialized, the accelerating pace of industry evolution across manufacturing generations amplified the role of direct, face-to-face social interaction among industry people for information exchange, learning, and knowledge-creation processes. The increasing importance of direct interaction, in turn, increased the value to companies of locating in proximity to one another in Japan. The importance of face-to-face contact among people and the accompanying benefits of proximity among industry members grew out of three important dynamics of the knowledge-creation process that we believe are general to any fast-emerging industry. These dynamics concerned the roles of individuals' tacit knowledge and unwritten understandings held within groups of scientists, engineers, operators, and managers; the rate and primary language of company and industry knowledge codification; and the volume and unpredictability of information flows in the industry as a whole. We explain each of these factors below.

*First,* the rapid pace of change and knowledge accumulation in individuals and groups outstripped the conversion rate of tacit knowledge to explicit (codified) knowledge (see Figure 7.1). All knowledge has tacit aspects that evade explicit representation, but form vital components of individuals' understanding of phenomena.[4] Individuals may build on their tacit knowledge to create new, explicit knowledge. They may, through reflection, find ways to explicitly convey some part of their tacit knowledge through speech, writing, or other forms of systematic representation. Even if individuals cannot explicitly represent many aspects of their tacit knowledge, this knowledge may play important roles in team, company or industry knowledge-creation processes in which people directly interact and can demonstrate to one another. Some of these interactions will help

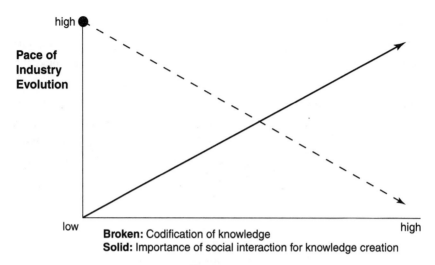

**Figure 7.1**

to make individuals' tacit knowledge explicit, and groups may arrive at shared understandings.[5] But even if this explicit knowledge becomes written, otherwise formally represented in blueprints or software, or embodied in equipment or materials, interpersonal contact remains essential to its full transmission.

No one ever succeeds in expressing everything they know about a topic, even in the most hospitable circumstances. In difficult circumstances, including rapid evolutionary change and information overload, individuals may rush communications, resort to shorthand forms of expression, rely more heavily on personal intuition, and miss connections with colleagues. They may spend more time taking personal action and less time reflecting on how to make their knowledge explicit for others.[6] If overall knowledge of a phenomenon evolves at an increasing rate, the proportion of knowledge about that phenomenon that remains tacit at any given moment may increase. If the tacit proportion of knowledge increases, the importance of interpersonal interactions in adding to knowledge also increases. As FPD industry evolution accelerated, individuals and groups acted as principal repositories of an increasing proportion of overall knowledge vital to companies' ongoing operations and progress.

In order to manage generational transitions, companies needed to develop processes to tap into knowledge embedded in individuals and groups. Consider, for example, NEC's efforts to barcode cleanroom operators and catalogue their behaviors when bringing early fab lines up from low to high yields. Companies found that generational transitions in manufacturing equipment required substantial numbers of experienced managers, engineers, and operators to be reassigned from previous generation lines to the new ones. At the same time, some proportion of experienced personnel needed to remain behind on older lines to keep them running at full yield and to train their cohort's replacements. As the time between generational changes grew shorter, some companies adopted new generation manufacturing lines before they had fully settled in predecessor lines. This negatively affected yields on the existing lines and retarded progress in waking up the new generation lines. Other companies skipped entire generations of FPD manufacturing equipment and substrate sizes in part because they could not staff the experienced engineer and operator hours necessary to awaken new lines without degrading productivity on the old ones.

*Second,* as the pace of overall knowledge accumulation intensified, the proportion of explicit knowledge that was written or otherwise recorded decreased in companies and the industry as a whole. Verbal discussions of R&D, process evolution, product needs, market opportunities, business outcomes, and technical standards raced ahead while systems to track, retain, and broadly transmit knowledge created in this way lagged. As a result the proportion of written, systematized information and knowledge at any given moment decreased while overall quantities of information and knowledge exponentially increased. The explosion in of explicit, but purely verbal information and knowledge challenged industry members to support broad intracompany and intercompany personal contacts to stay abreast while pushing technology, their businesses, and the industry ahead.

Industry life exhibited a surprising degree of collegiality among competitors as a consequence, including members of the international technical community, business managers in Japan, and senior international managers with connections in Japan. At the same time trade associations and government agencies accustomed to gathering and organizing information to promote cooperation found themselves unable to track industry change because they were not intimately engaged on the shop floors and in the company conference rooms where knowledge was created. Within companies, domestic and international movements of individuals and teams played a vital role. Companies and researchers lacking

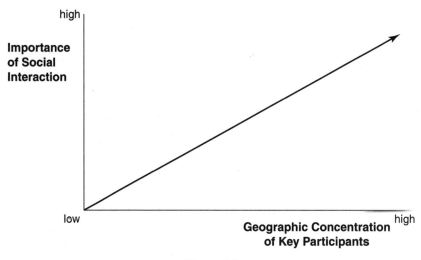

**Importance
of Social
Interaction**

low

**Geographic Concentration
of Key Participants**

high

**Figure 7.2**

strong Japanese connections sometimes reinvented the wheel, solving technical problems for which viable solutions already existed in the open market.

The rapid growth and concentration of the industry in Japan (see Figure 7.2) intensified the challenge for non-Japanese participants. Many important materials were written in Japanese, including manufacturing equipment manuals, scientific papers, patents, trade group reports, market research studies, and the most comprehensive current business media. Some of these materials were translated. Others were not. Even when materials were translated, the translation and dissemination process caused delays of up to a year in availability to non-Japanese speakers. As time went on, the infrastructure for cross-language information and knowledge dissemination improved. Consulting companies formed relationships with media, trade groups, and government agencies that gradually created a closer degree of convergence in the publication timing of Japanese and English language versions of important materials.

The immediacy of telecommunications, teleconferences, email, and real-time information systems like DTI's D-TIMES helped bridge the codification, distance, and language barriers. But these remote forms of communication proved insufficient by themselves. Successful non-Japanese industry participants found ongoing, live, one-on-one contacts and group discussions with Japanese participants essential to progress. These required an active Japanese presence

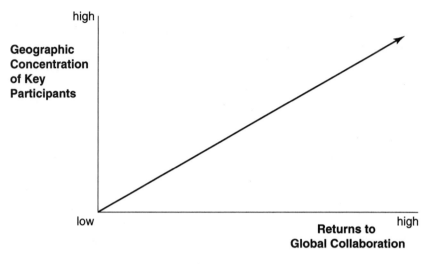

**Figure 7.3**

and rich, ongoing, collegial relationships with Japanese counterparts. Strategic alliances and/or particularly intimate collaborative relationships with affiliates, customers, and suppliers in Japan played important roles in successful industry entry by several non-Japanese companies, including IBM, Applied Materials, Philips, and Corning (see Figure 7.3). Compared to North American and European firms, Korea's geographic proximity to Japan, cultural similarities and history of Japanese colonization often worked to Korean firms' advantages as they worked to establish high-volume production beginning in the mid-1990s (see Chapter 8). Many Korean engineers and managers found they could understand Japanese technical manuals "even without knowing Japanese."[7] Airlines profited from the dense traffic of industry people moving in both directions on the short air hops between Kimpo, Narita, and Kansai International Airports. But the language gap and physical distance continued during the mid-1990s to cloud the understanding of many industry observers and participants—even in Japan—as the scale, timing, and early success rate of TFT LCD production start-ups in Korea and later Taiwan caught many by surprise.

The *third* dynamic affecting the importance of direct social interaction for knowledge creation in the industry concerned the volume and unpredictability of information flows. Random events critically influence the evolution of new businesses and industries, particularly early in their histories.[8] In Corning's case, for example, the unexpected arrival of important information in the form of unpredicted orders from large Japanese corporate labs for thin-glass products triggered a knowledge-creation process that eventually led to the establishment of a new advanced display materials business within the company. Managers never know when an event will occur that imparts information of such critical import. Managers cannot calculate the probabilities that such an event will occur at all. If and when such an event occurs, its importance will almost certainly not appear obvious, but may become clear only after considerable time has passed.

The fact that random, ambiguous events can change a company's history does not mean that managers should adopt a reactive mode. An event acquires meaning only if a company has the right people in the right place at the right time to notice it, interpret it, learn from it, and leverage it with other information. This means that continuity plays a critical role. Important events relevant to particular industries are more likely to occur in some places than in others, particularly when an industry is geographically concentrated as the mid-1990's FPD industry was in Japan. Companies with operations there enjoyed increased probabilities of coming into contact with such events. But the probability that a company would ultimately profit from such events equaled zero, unless the it implemented strategy processes that first, sensitized managers to potential opportunities in chance happenings; second, rewarded continuous engagement not just in gathering information but in transforming it into knowledge about opportunities; and third, provided incentives, responsibility, and authority to translate this knowledge into action.

We believe that by the early 1980s the high proportion of information-generating and knowledge-creating activities taking place in Japan around LCDs had significantly, positively influenced the probabilities that critical future events that pertained to FPD industry emergence would take place there. These probabilities increased with each new advance and with every new company that entered the industry in Japan, increasing the momentum of TFT LCD technology development relative to alternative technologies.[9] Most successful non-Japanese entrants recognized this, contributed to it, and maximized their chances of success by siting senior managers and researchers there on a continuing basis and giving them authority to act on what they learned.

## MAPPING THE DYNAMICS OF KNOWLEDGE FORMATION
## IN NEW INDUSTRY CREATION

As companies struggled to learn to manufacture the new, large-format TFT LCDs, they needed to prove a first generation of products, manufacturing equipment, materials, and processes and almost simultaneously prepare a new generation to succeed it. In the industry dynamic that emerged, new knowledge formed the basis of newer knowledge at an unprecedented rate of speed.

Figure 7.4 illustrates a dynamic process of FPD industry knowledge accumulation as it pertains to transitions from one product generation to another. Each of the four quadrants represents a phase in a single cycle of knowledge creation bounded by current and next generations. The spiral represents progress from one phase to another. We would chart an increasing pace of industry evolution as successive, tighter spirals.[10] As the pace of evolution amplifies, phases of successive generations increasingly overlap. In order to sustain a leadership position for their companies, managers, researchers, engineers, and operators must pay simultaneous attention to different phases of multiple generational transitions.

The spiral for a new FPD generation began as soon as companies integrated equipment and materials and woke up the first fabrication lines for its predecessor, or current generation (*Phase I/Southwest Quadrant*). At this stage, a significant proportion of current generation knowledge was embodied in FPD products, manufacturing equipment, and materials, and/or codified in patents, equipment manuals, and process recipes. These embodiment and codification processes, however, did not exhaustively, formally represent the knowledge created in the course of the current generation's commercialization. Significant aspects of current generation learning remained in the form of tacit and verbally transmitted explicit knowledge that researchers, engineers, operators, and business managers could utilize only individually or as interacting members of a social community (*Phase II/Northwest Quadrant*). Furthermore, current products, production equipment, and materials continued to support new knowledge formation from the current generation as teams integrated the new equipment, experimented to increase yields, and accumulated service and performance experience, records, and process manuals. Yet, even as participants put some of this knowledge to work optimizing current generation production lines, much of it functioned to create the foundation for the next generation. As generational life

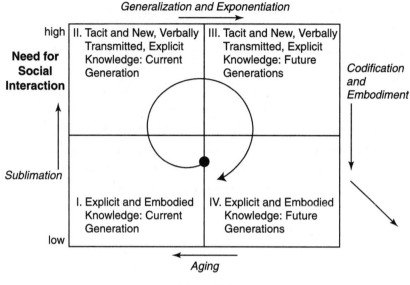

**Figure 7.4**

cycles grew more compressed, the proportion of new knowledge feeding directly into next generation rather than current generation activities increased.

Some of this knowledge short-circuited the spiral and became directly, immediately written, otherwise recorded and/or embodied in new generation projects (*Phase IV/Southeast Quadrant*). Most of it served as the tacit basis for generalizations that reassigned researchers, engineers, operators, and business managers carried forward to exponentiate the knowledge accumulation process around new generation projects (*Phase III/Northeast Quadrant*). Eventually, the experienced participants joined together with less-experienced participants to carry out the generational shift. In FPD industry evolution, generational transitions relied for success on the availability of sufficient numbers of experienced participants for reassignment, leveraged by increasing automation as generations advanced.

Generational shifts formally began as companies formed new teams to set about the work of designing new products, manufacturing equipment, processes, and business models (*Phase III/Northeast Quadrant*). Team members developed their initial thinking as individuals, and the knowledge they created in this fashion remained tacit at first. These tacit foundations formed the basis for explicit knowledge in both individual and group processes. Individuals from various companies met in informal networks and at professional meetings to discuss new generation prospects and to present papers on technical, business, and strategic issues. Team members created papers, proposals, and business plans that circulated both within their companies and outside their companies as important inputs to supply agreements, sales contracts, downstream product development processes, and discussions of industry technical standards.

Meetings intensified, at first to share ideas and soon to push for results, such as prototypes, that began to embody the new generation knowledge (*Phase IV/ Southeast Quadrant*). Producers, equipment suppliers, and materials suppliers set objectives to move as rapidly as possible to the phase of knowledge codified in manuals and process recipes, and embodied in the form of equipment, materials, facilities, and products. The pressure built within producers to ramp-up for full production before competitors did. Once any company achieved commercial yields, the future generation rapidly aged to become the current generation. As in previous generations, some proportion of the knowledge carried through the process remained tacit and some proportion of explicit knowledge remained unwritten, unrecorded, unembodied.

In FPDs, the accelerating pace of industry evolution amplified the role of direct social interaction among people in learning and knowledge-creation processes. The role of tacit and verbally transmitted information and knowledge in learning and new knowledge creation also transformed in magnitude, character, and scope. In the visual language of Figure 7.4, knowledge spirals became shorter, tighter, asymmetric, and also more elliptical as the proportion of information and knowledge that was written down or otherwise systematized at any given moment decreased, while the overall quantity increased. Successive generations incorporated increasingly recent experience from prior generations to produce the tacit and explicit knowledge basis of immediate subsequent developments. Companies needed geographic proximity to each other and community membership to sustain these interactions. Without Japanese operations and proximity to one another, companies could neither encounter nor interpret principal information and knowledge flows that were driving the technology and the industry forward.

We believe that interaction among learning, continuity, speed of industry evolution, the character of knowledge accumulation, and geographic concentration represents a general phenomenon in new industry creation. The increasing pace of industry evolution causes an increase in the proportion of tacit and verbally transmitted explicit knowledge, which in turn raises the importance of direct social interaction for knowledge creation. This sets the stage for a development phase in which geographic concentration of participants' new physical locations first becomes desirable and then autocatalytic or self-reinforcing. The need for speed creates a premium on proximity among producers as well as their materials and equipment suppliers.

The need for speed also counterbalances companies' interests in closely holding proprietary information that pertains to their competitive advantages in manufacturing current products. Strategic alliances and close relationships with suppliers and customers involve significant disclosures. Companies focus on winning the race to successfully manufacture the next generation, rather than on holding on longest to advantages built on the previous generation. They compete on knowledge creation rather than product positioning.

## FREEING AFFILIATES FOR CREATIVITY

The high-volume FPD industry did not emerge solely as a consequence of technological innovations, but equally as a consequence of strategic and organizational innovations. In order to succeed, managers needed to understand the relationships between global and local phenomena in a new way. Companies incorporated these new understandings into strategy and organization by re-examining old, hierarchical assumptions about headquarters/international affiliate relations.[11] Some delegated business leadership to affiliates outside of their home countries. Others mandated global production responsibility or global oversight of other functions for particular businesses to affiliates. Many increased R&D outside of their home countries. Companies that took these steps found them necessary to build and sustain advantages in a market where competition was born global.

The strategic and organizational innovations we observed in FPD industry creation hold general significance for many industries. The idea of truly devolving global responsibility and authority—including technology and new business creation—to units outside companies' home countries remains controversial in most organizations. Many international companies have built world-class organizational capabilities in multiple countries, but continue to rely exclusively on home-based resources as platforms for creating new knowledge. Foreign R&D operations exist primarily to adapt existing products to local markets and keep an eye on competitors' technologies. Technology dependence serves as an instrument of strategic control in many companies. Affiliates, the thinking goes, face incentives to coordinate strategy within the international firm to the extent that they must rely on home country R&D units to advance their products and technologies. If affiliates gain the technological initiative, they may gain too much autonomy and pursue their own interests at the expense of corporate interests.

Global high-technology competition demands a more flexible approach. Companies need strategies that focus their full global capabilities on new industry opportunities, because the speed of technological advance strains capabilities to master one product generation before another comes along. Because location can confer advantages early in the evolution of a new high-technology industry, companies can gain these advantages and avoid overlooking opportunities by decentralizing new industry initiatives to affiliates. This does not mean that companies should cut their international affiliates loose to pursue independent initiatives. Decentralizing initiatives means expecting international affiliates to take the lead as new industries emerge in their countries by leveraging their proximity and country-specific capabilities with corporate knowledge and capabilities from throughout the world.

## LEVERAGING VALUE ACROSS GEOGRAPHY, PARTNERSHIPS, AND TIME

The strategies needed to manage knowledge creation in new, global high-technology industries contrast with lessons companies learned as industry competition increasingly internationalized in the 1970s and 1980s. Initially, companies internationalized their strategies to exploit firm-specific advantages by leveraging their value chain activities across geographic locations to create cost advantages, differentiation advantages, or both. As the pace of competition and technology evolution grew more heated, companies increasingly leveraged their activities with the competitive advantages of alliance partners, suppliers, and customers to gain benefits of collaboration.[12]

Competing in knowledge-driven industries challenges managers to do more than think about leveraging value chain activities within their companies, across geographic space, and among allied organizations. Knowledge-driven competition challenges companies to leverage their value chains in time, across new technologies and technology generations.

Success requires a competitive orientation that encompasses the traditional, product-driven generic strategies of cost leadership and product differentiation with a new, dynamic value proposition: speed. In the remaining sections of this chapter, we contrast attributes of the traditional, product-driven generic strategies —cost leadership and differentiation—with critical attributes of a global, knowledge-driven strategic orientation toward new industry creation. Table 7.1, which unpacks the ideas presented in Table 1.1, summarizes our argument.

### A New Value Proposition

Companies compete by creating and sustaining firm-specific competencies that produce benefits that customers value.[13] The value proposition for cost-driven strategies demands that companies produce their products and services at lowest cost relative to competitors. Companies that meet this objective can profitably undercut market prices to attract customers away from competitors. Differentiation strategies operate on a value proposition of perceived or actual product/service uniqueness. Companies establish uniqueness in a variety of ways. Examples include specializing within a line of products and services to meet the needs of different market segments, establishing brand names and attributes, establishing reputations for exceptional quality, or bundling offerings in innovative ways, as in the case of wireless telephone service, computer software, Internet access, and technical support. Companies that pursue differentiation strategies can command higher margins because customers do not perceive competitors' offerings as direct substitutes and willingly pay premium prices for product attributes they perceive as specially suited to their needs.[14]

Knowledge-driven strategies address a crucial vulnerability of every profit source: the passage of time. Profits from differentiation and cost-driven strategies inevitably erode as competitors imitate market leaders' innovations and improve

**Table 7.1** Principal Attributes of a Knowledge-Driven Competitive Orientation Compared to Generic Product Positioning Strategies of Cost Leadership and Differentiation

| Attribute | Competitive Orientation | | |
| --- | --- | --- | --- |
| | COST-DRIVEN | DIFFERENTIATION | KNOWLEDGE-DRIVEN |
| Value proposition | Cost leadership | Uniqueness | Speed |
| Competitive advantage | Products and processes | Products and services | Competencies and functionalities |
| Internationalization: R&D process | Efficiency-seeking: central R&D creates standardized global products | Market-seeking: local R&D adapts global products for local markets | Knowledge-seeking |
| Internationalization: new business development process | Global leadership from home country | Affiliates choose opportunities from central repertoire | Affiliates and center lead and leverage each other globally |
| Sustaining competitive advantages | Protect, exploit, extend | Protect, exploit, adapt | Create, share, transcend: drive for next generation |
| International integration | Vertical supply autonomy; selective alliances and purchase contracts to redress scale disadvantages | Captive suppliers of critical value chain activities; selective alliances and purchase contracts to redress competence gaps | Alliance, supplier and merchant relationships for key activities to learn, enhance speed to market, and secure location advantages |
| Intrafirm network | Center-outward | Center-outward and inward | Multiplex |
| Globalization | Project, protect home national positions | Project, adapt home national positions | Leverage multiple national strengths |
| Nationality | Dominance | Co-option | Collaboration |

their manufacturing processes to undercut the leaders' costs. Innovators face a finite window of high profitability for any new product: the time period for before imitators arrive on the scene and price competition sets in. Knowledge-driven competition narrows that window.

Knowledge-driven strategies set an objective of speed to stay ahead of competitors in delivering a stream of new products at decreasing cost. Continuous product innovation props open the window of profitability as old products fade, draw imitators, or fall into commodity patterns of competition. Continuous process innovation couples product advances with advances in manufacturing efficiency. In order to achieve profitability as the large-format, high-volume

FPD industry emerged, companies needed to rapidly, simultaneously expand the frontiers of product and manufacturing innovation as measured by increased substrate sizes, throughput, and manufacturing yields. Successive generations of new products arrived on the scene coupled with new manufacturing equipment, materials, and techniques that continually drove costs down.

### Knowledge-Based Competitive Advantages

Knowledge-driven strategies require managers to broaden their perspectives beyond products and processes as sources of competitive advantage to consider core competencies and functionalities. Gary Hamel and C.K. Prahalad, who originated the idea, defined core competence as a bundle of skills and technologies that together deliver a customer benefit.[15] Viewing strategy as a function of core competencies liberates capabilities from the products and organizational units in which they are embedded, so companies can rapidly recombine and redeploy them for future product/service opportunities. Functionality thinking similarly liberates user benefits from end product and service embodiments, encouraging flexibility and innovation in designing solutions to meet particular customer needs.

The FPD industry of the early 1970s exemplified well the differences in competitive outcomes that can emerge between companies that espouse core competence mentalities as opposed to those in which current market orientations hold sway. Some companies invested in delivering light, portable, visual functionalities that had the potential to translate into customer benefits in a variety of contemporary and future end products. The underlying core competencies for calculators, watches, and later small TVs also created the foundation for notebook computer screens, monitors, and eventually wall-hanging TVs. Many other companies, particularly in the U.S., invested in FPD technology with the hope of moving directly to giant, wall-hanging TV monitors. When these companies abandoned the TV market, they not only stopped making traditional CRT television monitors, but also stopped their FPD research programs. In doing so, they severed underlying technologies and skills that might have formed the basis for future competencies. Some companies, such as Lucent and Motorola, tried unsuccessfully to build these competencies again from scratch in the 1990s.

### Internationalization of R&D and New Business Development Processes

Knowledge-driven strategy implementation relies on maintaining global innovation capabilities in R&D centers in multiple countries. But establishing such centers will not, in itself, establish a knowledge-driven strategy. The linkages and processes that connect R&D, corporate, and business leadership within and across countries play a critical role. Many companies establish R&D centers in locations where they can take advantage of rich institutional contexts for innovation—for example, proximity to research universities, access to government programs and consortia—as well as labor markets abundant in educated, experienced research workers.[16] Often, however, these operations function as ivory towers or acolytes of larger laboratories at home. Isolated from business and corporate top management, these labs' potentials are as remote from business leaders' attention as the leaders are geographically remote from the labs. Even if top managers in many companies view offshore R&D centers as augmenters of home-base advantages rather than adapters, they rarely view them as originators of global advantages for the company as a whole.

Knowledge-driven strategies do not predispose new business development leadership to co-locate in accordance with pre-existing geographic value chain configurations, business headquarters or pecking orders within the technical community. Home and local affiliates alike actively identify new business opportunities that have global potential. All affiliates give attention to the opportunities identified by foreign counterparts as well as to their own advocacies. Companies assign leadership for new global businesses to affiliates in countries where the technical foundations and market opportunities are emerging most rapidly. These affiliates hold the best positions to identify, understand, and advocate these opportunities' potential for the corporation and to globally leverage country-specific advantages, the company's core competencies, and other affiliates' knowledge to pursue them. Unless companies take rapid R&D and new business development initiatives at the affiliate level, many new industry opportunities that emerge outside of their home countries fade into the past before the relevant managers even notice them.

Knowledge-driven orientations toward R&D internationalization and new business development contrast with cost- and differentiation-driven strategies, both of which tend to centralize the critical aspects of these activities at home.

Cost-driven strategies centralize R&D to leverage new technology platforms across product divisions to create standardized products for global markets, as well as capture economies of scope in cost-saving manufacturing process improvements. Home country operatives lead product and new business development on a global basis. Differentiation strategies tend to concentrate power at home to create new core technologies and control new technology applications. Smaller R&D operations outside of the home country often adapt core technologies and products to local preferences. New business development leadership resides in the home country. Affiliates generally choose new businesses for their local markets from companies' global product portfolios. In some industries, such as packaged goods, affiliates may also have the discretion to establish and market local brands.

The most successful U.S. companies in the emerging FPD industry—in particular, IBM, Applied Materials, and Corning—recognized the knowledge-driven nature of competition in the new industry and assigned responsibility and authority for new business development to their Japanese affiliates. Intensive communications and personnel exchanges helped these affiliates, in turn, to leverage and enhance their R&D capabilities as peers and leaders within the corporations' global technology and marketing networks. Senior managers recognized that the geographic concentration that accompanied the early phases of the industry's emergence represented fertile ground for growing new businesses. In such circumstances, it is risky to uproot new business ideas and transplant them to headquarters. Companies avoided this risk by creating new, more flexible processes for global new business development and cross-fertilization. Corning did just this when senior managers transferred Satoshi Furuyama from Japan to Corning, New York, to work with them to develop the substrates business and work on a strategic plan, and then back to Japan to implement the plan as president of Corning Japan K.K.

### Key Processes to Sustain Knowledge-Based Advantages

Managers implementing traditional cost-driven and differentiation strategies accept as gospel the need to hold critical knowledge resources closely to avoid diffusing them to competitors and potential competitors. This "need to know" orientation toward knowledge-sharing applies not only to relationships with competitors, suppliers, and customers but also with affiliates outside of the home country. These affiliates typically find themselves veiled off from the key organizational processes that create and sustain knowledge-based competitive advantages.

Cost-driven strategies exploit new advantages by embedding them in globally standardized products with manufacturing process recipes designed mainly at home. They sustain the profit streams from innovations for as long as possible by relying on continuing process improvements to drive costs down.[17] Cost-driven strategies tend to assign downstream activities with intrinsically local dimensions, such as sales and marketing, to companies' international affiliates. If affiliates expand into manufacturing, they often perform generic activities such as assembling intermediate or final goods from components produced elsewhere. Such activities do not require the company to transfer in core knowledge. As companies expand, some international affiliates may gain strategic production responsibilities, such as regional mandates or global responsibility for sourcing of key (usually locally abundant) inputs. But they remain on the periphery of the knowledge-creation processes that sustain the company's competitive advantages over the long term.

When national preferences define significant product differentiation opportunities, international affiliates have opportunities to increase the salience of their roles in creating and sustaining their companies' knowledge-based advantages. Many contribute by adapting their companies' core technologies and products through line extensions, product improvements, or other alterations to suit national or regional tastes.[18] Affiliates' adaptive innovations may have global potential or appeal to market segments in other countries with similar needs or tastes. Companies face a challenge to find processes that amplify the significance of these contributions beyond the borders of the countries where they originate.

The few companies that meet this challenge on a regular basis have done so by globalizing affiliates' new business ideas from the center. This means assigning responsibilities for exploiting new ideas with global potential back to worldwide product management teams based at home. In the FPD industry, 3M took this approach with its highly successful TFT LCD brightness enhancement film. Sumitomo 3M, the company's Japanese joint venture affiliate, identified the market opportunity as an application of 3M's microreplication technology. Business leadership as well as R&D, however, remain centered at 3M headquarters in St. Paul, Minnesota. Control remains in St. Paul, in part, to assure that the company can leverage continuing advances in microreplication technology across multiple product markets, and in part out of concern that knowledge diffusion will raise up imitators.[19]

We believe that the future pace of high-technology competition will increase pressure for more affiliate-based approaches. The need for speed counterbalances the imperative to internalize and protect a technology at home. Knowledge-driven companies sustain their knowledge-based advantages by rapidly transcending current products and creating their successors. By the time an imitator moves into a market, companies must have moved on to the next generation, rendering current products obsolete.

The U.S. companies that succeeded in the FPD industry during its take-off period in the early 1990s delegated responsibility and authority for running new global businesses, once established, to their Japanese affiliates and/or alliance ventures. We believe this step was necessary in part because emerging high-technology industries present a continuous stream of new concepts—like those underlying rapid generation changes in TFT LCD manufacturing technology—that rapidly build upon and transform each other. Effective participation in such environments requires companies to empower local affiliate managers not only to observe events and be heard, but also to act and to lead.

### Orientation Toward International Integration

Since at least the mid-1970s, multinational corporations (MNCs) have dominated international trade in products[20] by managing flows of firm-specific resources among integrated global networks of owned affiliates. Companies apportioned their value chain activities among countries to exploit cost-reduction and differentiation opportunities, integrating cross-nationally to create internalized production and trading networks. Managers based strategy implementation on ownership of key activities to reduce risks that core technology, brand equity, or other advantages would diffuse to trading partners who might later rise up as competitors.[21] In a high-speed world economy dominated by trade in knowledge, companies need more flexible strategies that permit them to mutually invest firm-specific resources in open, global, knowledge-creation processes within networks of alliance partners, customers, and suppliers.

Cost-driven strategies have traditionally relied on cross-national vertical integration. Some companies centralized most of their value chain activities in one or a few countries and transferred the final goods to marketing affiliates around the world. Toyota followed this approach until the early 1980s, performing virtually all of its design and manufacturing activities in Japan. Alternatively, companies implemented vertical integration in a decentralized fashion by assigning global mandates for specific value chain activities or complete value chains for particular products to relatively large numbers of affiliates. The affiliates then transshipped the resulting parts and products among themselves to complete their companies' lines in every country where they operated.

Highly integrated strategies and organization structures biased managers' decision-making toward self-supply. Companies entered purchase agreements or alliances selectively to redress scale diseconomies or gaps in market knowledge. Companies often used these relationships as learning opportunities to prepare for acquisitions or the start up of new, wholly owned national affiliates.[22]

Companies used either vertical or horizontal integration to pursue differentiation strategies. The importance of market responsiveness under differentiation, however, created incentives for decentralized approaches that distribute production among as many countries as efficiently possible to increase proximity between high value-adding activities and customers. Horizontally integrated companies established equivalent, relatively complete value chains for all of their products in almost every country where they operated, creating miniature versions of their home country operations.

The uniqueness claims that form the basis of differentiation strategies motivate companies to own and control the value chain activities that embody their advantages. Companies pursuing differentiation strategies within the same industry, consequently, often differ in their integration patterns. The U.S. systems integrator Dell, for example, externally sources all of the components for its notebook computers, including FPDs. It assembles every computer it sells to order, as specified by individual customers. Customers order directly from the company. Logistics competencies and capabilities in order generation and fulfillment make Dell's success possible,[23] and the company integrates these activities. Other notebook makers, including NEC and Samsung, regard display attributes as major factors that differentiate their products. Consequently, they identified FPDs as strategic components and integrated backward to produce them. These companies continue to compete vigorously with one another to advance FPD technology in such areas as size, brightness, resolution, and viewing angle. Conventionally, companies pursuing differentiation strategies enter purchase agreements and alliances only to redress competence gaps or to obtain materials that do not affect their advantages over other producers. All notebook producers, for example, rely on outside microprocessor producers such as Intel or AMD.

Knowledge-driven strategies challenge managers to formulate more flexible approaches to integration that leverage potential contributions to competitive advantage of strategic alliances and close customer/supplier relationships. FPD development histories at Sharp, Toshiba, and IBM illustrate increasing returns to these relationships as the pace and geographic concentration of industry evolution increased.

In the early 1970s, researchers at Sharp and Seiko Epson foresaw LCDs' potential to enable a broad range of yet-to-be-imagined products and led the way to commercialize the technology. Once they achieved their breakthroughs, they opened up component markets for the new LCDs in addition to incorporating them in their own products. The decisions to share the breakthrough technology with other companies through component sales helped develop the market, benefiting both Sharp and its competitors. Sharp achieved scale and learning economies more rapidly than they might otherwise have done. Merchant sales helped to fund continuing R&D, which enabled Sharp to introduce a succession of products based on continually improving display functionalities, including small televisions, personal digital assistants (PDAs), video cameras with FPD viewfinders, digital cameras, and the first complete line of flat home TVs. Customers for Sharp's LCDs engaged their own distinctive strengths and brand names to multiply successful new applications, such as the Sony Video Walkman. The increased pace of market development also created positive feedback effects in equipment and materials.

Toshiba and IBM seized leadership positions in the second wave of companies to grasp the technology's significance and acted on senior managers' insights by forming the DTI strategic alliance. Despite Sharp's overall FPD lead, during the mid-1990s DTI held the dominant market share in the most advanced, large-format TFT LCDs used in high-end notebook computers. IBM had focused early on screen size and color capabilities as critical attributes to differentiate the company's notebooks, and they built significant basic FPD research and R&D capabilities. Toshiba had established itself as a manufacturing process leader.

The DTI alliance permitted the two companies to combine their resources, learn rapidly, and dramatically cut the time to market for their initial TFT LCD offering. By awarding a leadership mandate to IBM Japan, IBM tapped into location-specific factors contributing to industry emergence, particularly social and professional networks among researchers in Japan. DTI contributed to the leading positions each parent enjoyed in high-end notebooks beginning in the early 1990s, although neither IBM nor Toshiba restricted its notebook businesses to DTI FPDs. The partners also sold a large proportion of DTI's output to outside customers. DTI's presence in Japan and its sales in the open market enhanced its learning capabilities and therefore the amount, value, and potential of the accumulated knowledge the alliance represented to its owners. For IBM and Toshiba, DTI represented a shared competence that nonetheless played a key role in their individual strategies to compete with each other and with other manufacturers by differentiating their high-end notebook lines. But the partners established the alliance as an investment in learning to stay ahead of future markets for products yet to be designed.

### Intrafirm Network Orientation, Globalization, and Nationality

In order to establish itself as a global player in an emerging high-technology industry, a company must compete and learn in the most dynamic possible industry setting: the geographic region where the industry is evolving most rapidly. Competing in the cluster exposes the company to the toughest possible rivals. This means fighting for more than market share. The cluster provides a critical learning context, incorporating professional, social, and institutional networks that offer the richest, most up-to-date news, information, and unwritten knowledge about a technology and industry. Competing in the cluster means competing for the most knowledgeable partners, suppliers, customers, and allies.

Competing globally with home-created, home-based advantages has built many of the world's greatest companies. Most companies that pursue cost- and differentiation-driven strategies honed their international competitive advantages within the supporting infrastructure of their home countries, competing with rivals there. They embodied their advantages in products and technologies that they projected outward from headquarters to affiliates around the world. In the simplest of such arrangements, affiliates interacted with headquarters, but rarely, if ever, with each other. In more complex arrangements, headquarters mediated interactions among affiliates, for example, to share information on best managerial practice. But the sources of competitive advantage remained home-based.

Managing a company's evolution from a simple cost- or differentiation-driven strategy to a knowledge-driven strategy requires both historical perspective and foresight. For many companies, competing internationally on purely home-based competitive advantages has already taken on the character of an evolutionary phase slipping rapidly into the past. This change does not reflect unrelated diversification, but a fundamental shift in the nature of markets brought on by the spread of high technology and digitalization. Many products morph beyond companies' traditional home-based competencies. How does a traditional supplier of paper family photographic albums, for example, come to terms with the advent of digital cameras and laser disc storage? Digital convergence creates product interdependencies, such as between computing, wireless telephony, video, and Internet access—where none existed before. Important industries have grown up around components—such as FPDs—that some managers once regarded as generic commodities. Changes such as these occur rapidly. They alter the set of relevant technologies, customers, suppliers, partners, allies, and competitors that a company must understand and face. In order to leverage home-based advantages and create new advantages in the face of these changes, most companies will need to forge new ties to new partners in new places.

While cost- and differentiation-driven strategies *project* competitive advantage, knowledge-driven strategies also *assemble* competitive advantage by discovering, combining, and recombining both home- and affiliate-based resources. Knowledge strategy implementation depends on developing an organization structure that operates with multiple centers of authority and responsibility to manage multilateral flows of knowledge, people, funding, and technology among affiliates in all countries. These flows, combined with strong ties to alliance partners, customers, and suppliers, play critical roles in internationally leveraging the distributed competencies that form the competitive strength of the knowledge-driven company.

## BEYOND LOCATION ADVANTAGE

International high-technology competition has recast the balance between companies' needs to protect the knowledge that underlies their competitive advantages and the need to continually learn more. Companies have historically built the foundations of competitive advantage at home and competed internationally by projecting that advantage outward into the world. In new, high-technology industries, companies must build the foundations of competitive advantage in multiple countries, wherever in the world knowledge is evolving most rapidly. For companies to successfully participate in creating the high-volume FPD industry, this translated into a need to develop strong bonds of both competition and cooperation in Japan.

We have argued that the FPD industry concentrated in Japan as it emerged because knowledge of the technology accumulated there more rapidly than anywhere else in the world. We have seen how the pace of industry change, the role of unwritten knowledge embedded in individuals and groups, and the foreignness of the environment to western firms meant that much of the knowledge that accumulated at first did not travel easily. U.S. and European companies needed an established presence in Japan to participate on an equal footing in the industry-wide learning process. Both U.S. and Japanese companies benefited from this co-location. We can fairly state that the fast pace of industry evolution was set in Japan, but Japanese firms did not set it alone.

Japanese presence by itself cannot explain the success of non-Japanese firms in the FPD industry. Companies needed strategies and processes that provided their Japanese affiliates with a high degree of authority and responsibility to create, develop, and globally lead their FPD businesses. They also needed relatively open learning regimes involving suppliers, customers, and alliance partners. Without such processes, knowledge accumulation would have quickly, permanently outrun them, as it has threatened to do to most firms in the industry at one time or another. If companies in other industries were to adopt such processes, we believe that most would need to significantly reorient managerial attitudes toward the roles of foreign affiliates, suppliers, customers, and alliance partners. Nonetheless, we believe that many of the lessons learned by successful FPD industry participants generalize to companies competing in the early phases of many industries. We have provided several frameworks in the chapter to aid in this generalization.

In the next chapter, we will turn our focus to events outside of Japan during the 1990s, particularly in the United States, where government took a strong role relative to that of the private sector in promoting domestic FPD production, and in Korea, where the inverse held true. We will see how companies' decisions to embrace or resist the industry's core dynamic of global knowledge inter-dependence created different outcomes within these countries' national territories. We will also see how a European company positioned itself for global market leadership late in the 1990s—just as U.S. companies had done in the late 1980s—through an Asia-based strategy of collaboration.

# 8

## Knowledge and Nationality

*When you're not number one, you need to
learn from those who are ahead.*

KYE HWAN OH[1]

*It's very difficult for government to guide
industry. Besides, it's old-fashioned.*

JAE-CHOON LIM[2]

*You shouldn't be planning an addition to
your house when the damn thing's on fire.*

JAMES M. HURD[3]


**T**he new industries of the future have no nationality. Countries share the benefits of industry creation in proportion to their firms' access and openness to global knowledge-creation processes. Of course, facilities and people occupy physical spaces at any given moment, generally defined, at least in part, by national territories. But knowledge creation at the company and industry level depends on managing technology and market interdependencies that cut across firm and national boundaries. Companies navigate these interdependencies most successfully by decentralizing initiative, authority, and responsibility to affiliates and ultimately individuals to create and lead new global businesses. From the perspective of headquarters, less management achieves more.

Similar considerations apply to government policy. Given the importance of high technology to national security and economic growth, most governments consider it a responsibility to play roles in new industry creation. Developed country governments have often succeeded in this role. But politics always restricts the types of policies governments can effectively implement for these purposes.[4] The dynamism and intensity of knowledge-driven competition creates further limitations to effective government intervention, which we will discuss in this chapter. Effective public policy remains carefully circumscribed, acknowledging that even a marginal intervention to counter market forces in the fast-moving, chaotic early phases of a high-technology industry's evolution can bring about sweeping, unintended consequences that may prove very difficult to reverse.

When companies vied to establish FPD production in Korea and the United States during the 1990s, differences between the two countries in the roles and relationships of business and government played an important role in the outcomes. Japan's preeminence as a production platform attracted public policy-makers' concerns in both countries. In 1990, 90 percent of TFT LCD production took place in Japan.[5] The figure had grown to 94 percent by 1994, with less than less than 3 percent attributed to the United States.[6] U.S. companies' failures to establish commercial scale manufacturing began to arouse national security concerns, as FPDs were expected to play key roles in avionics, weapons, and battlefield logistics systems. In the late 1980s, the Department of Defense (DoD) began to offer significant funding aimed at improving FPD technology development for military applications. Funding continued through the 1990s in the form of various programs and initiatives, some aimed at creating interfirm and industry collaboration both domestically and internationally. The government of the Republic of Korea also undertook funding programs during this time period, but spent far less money relative to the private resources invested in the industry and achieved considerably less impact on corporate strategies and intercompany relationships. By 2000, more than 36 percent of global TFT LCD production took place on Korean soil,[7] while overall FPD production in the U.S. had declined to less than 1 percent.

## POLITICS AND NEW INDUSTRY CREATION

In this chapter, we will compare the 1990s strategies of Korean companies with those of U.S. firms that focused on U.S. production, drawing lessons for both governments and companies. Governments often offer themselves as midwives and strategic partners in new industry creation and may provide valuable assistance to companies in this capacity. Effective corporate strategy, however, particularly in high technology, demands sensitivity to the limits of government capabilities and to the potential constraints on managerial flexibility inherent in these offers. Despite public policy-makers' best efforts to the contrary, mixing the political process with the new business-creation process introduces non-market contingencies that can cloud the criteria of *learning, continuity,* and *speed* that we have described as vital elements of strategy in knowledge-driven competition.

## CONTINUITY, POLITICS, AND POLICY INCONSISTENCY

New, knowledge-driven industries do not emerge overnight, but reach critical mass as a culmination of enduring, continuous efforts on the parts of companies. Government programs to support basic research can play an important role in growing a public pool of broadly applicable knowledge from which many such private efforts may spring. As technologies underlying new industries evolve away from their basic research foundations toward specific products and markets, however, governments' capabilities to help decline. Implementing programs to promote specific technologies and product markets amounts to industrial targeting, which places a government in the position of substituting political judgments for market forces in picking winners and losers. Such bets carry incalculable odds early in an industry's evolution and may cause harm by favoring weak competitors over strong ones or otherwise distorting market incentives.

Outside of defense industries, most managers find government programs potentially beneficial on a tactical level, but too unreliable to form a basis for strategic decisions, such as whether or not to enter a new industry. As policy-makers increase the specificity of programs by targeting particular industries, technologies, or companies, the programs become more difficult to sustain in the face of objections from groups that do not benefit. Programs that appear to benefit narrow or politically unpopular interests can draw fire and may lose out to broader or more powerful groups. Government policies and government officials may also change due to elections, shifting interest coalitions, intragovernmental or interbranch rivalry and power shifts, or a failure to persuade all vital constituencies to cooperate. Complex programs require the support of diverse, overlapping, sometimes conflicting interests that may prove difficult to reconcile. Governments may try to circumvent the problems we have described here by offering programs that spread benefits widely across many technologies and companies. Such programs may carry high price tags and spread benefits too thinly to have much positive effect, while at the same time imposing conditions that restrict participants' strategic flexibility. Implementation may also suffer from an adverse selection problem—for example, weak competitors may apply in disproportionate numbers. Strong competitors value flexibility and base their strategies on business criteria that will hold up with or without a government's help.[8]

The U.S. government's involvement in programs to promote FPD production in the U.S. provided many examples of potentially disruptive policy discontinuities arising from considerations like those we have described here. The unexpected loss of democratic majorities in both houses of the U.S. Congress in November 1994, for example, set off a complex chain of events that severely hampered FPD industry plans that the Clinton administration had announced the previous April. Foreseeing the congressional gridlock that emerged in the wake of that election, the federal official most personally identified with the administration's National Flat Panel Display Initiative (NFPDI) left government the following spring to return to prior research interests, leaving the new program essentially leaderless.[9]

The idea of promoting a domestic FPD industry enjoyed bipartisan support, even in the new Congress. But one of the principal underlying funding sources, known as the Technology Reinvestment Project (TRP), was cut because it appeared to Republicans in Congress that it endowed the Democratic administration with discretionary resources to disburse for the president's political gain.[10] TRP guidelines, as well as DoD policy more generally, opened the program to foreign participants. If the proportion of TRP funding in the overall program had remained significant, as planned, U.S. public policy initiatives in the industry might have taken on a more internationally oriented character than turned out to be the case after 1994.[11]

## LEARNING, KNOWLEDGE, AND TERRITORIALITY

In order to achieve global competitiveness in new industries, companies need access to leading-edge technology and to the knowledge-creation processes that drive its evolution. Government programs to promote new industries can help companies to meet this need only if policy-makers turn an indifferent face to the national origins and geographic context of the forces driving industry advance. In practice, this degree of openness can prove very difficult to establish, particularly when a program arises from specific concerns about national competitiveness or military readiness and depends on tax funds for support.

Modern governments by definition serve as authorities within territorially defined jurisdictions and draw most of their financial resources from citizens. It makes sense for governments to expend resources in ways that benefit citizens and non-citizens alike, however, as long as inclusiveness benefits the country more than exclusivity. Of course, many government programs provide public benefits—roads, for example—from which it is neither practical nor desirable to exclude particular groups (such as non-citizens or tax avoiders). Much government support for basic research falls into this category. But as programs take on a more focused character by targeting industries, technologies, or companies, they create private benefits from which groups can be readily excluded. In debates over the distribution of costs and benefits of such programs, many politicians find the inclusion of foreign-owned beneficiaries, such as technology licensors or local affiliates of multinational corporations, difficult to justify to their constituencies. Issues of national autonomy also come into play, when an industry's products serve a national defense need, production is geographically concentrated, and foreign supplies do not appear forthcoming.

The politics of foreign participation in publicly funded industry promotion schemes can result in a complex, legalistic thicket of contradictory signals for corporate strategy. In the U.S., for example, the Defense Advanced Research Project Administration (DARPA) welcomed foreign participation in its FPD initiatives. Congressional authorization/appropriation language, however, placed restrictions on using TRP funding in this way. One key Department of Defense decision-maker referred to these restrictions as "minimal," and the program as fundamentally encouraging to foreign companies.[12] Yet many observers in business and other departments of government, including the State and Commerce Departments, regarded congressional mandates restricting foreign participation in federally funded technology programs as sufficiently ambiguous to impede funding of groups that included foreign partners.[13] While policy-makers contemplated a degree of openness that might, in principle, have funded a 100-percent foreign-owned company to produce FPDs on U.S. soil, DARPA program managers expressed reservations about the politics of actually doing so.[14] In the view of DoD policy-makers, companies working on DARPA-funded FPD production equipment contracts have always enjoyed flexibility to beta-site with any producer in the world. Yet in 1996, an industry official with responsibilities in a key DARPA-funded FPD industry promotion program stated that a foreign beta site, would "violate DARPA internal procedures. It would be out of the question."[15]

The issue for corporate strategy does not hang on the outcomes of such political debates, so much as on the fact that such debates take place. We have seen that knowledge-driven strategy rests on international knowledge exchange and the ability to partner with global technology leaders to create new knowledge. Managing the knowledge imperative requires companies to decentralize initiative so that individuals and their organizational subunits can engage actively in the knowledge-creation process whenever, wherever, and with whomever it is happening. Relying on government industry promotion programs as a core element in strategy detracts from the clarity of these business priorities. It also retards the autonomy of business decision-making by adding company-specific public policy contingencies and political risks[16] to the generic contingencies and risks of participating in new industry creation.

Public policy programs charged with a mission of building a national presence in an international industry place knowledge creation at risk in another fundamental way. As John Seely Brown and Paul Duguid have suggested, individuals' sense of self-identity influences what information they pay attention to, how they attribute meaning, and what they learn.[17] In new industry creation, we suggest that a corresponding sense of others' identities will also have an impact, particularly on the social context and partnerships perceived as relevant for learning and knowledge creation. Nationality represents a powerful dimension of individual identity that forcefully directs attention to a distinctive history, culture, economy, body of civic values, and set of technical achievements as foundations for progress. Identifying with a mission to create a national industry can easily bias managers to search locally rather than globally for relevant knowledge, knowledge-creation partnerships, and frameworks for industry understanding. As a consequence, optimal learning connections to customers, suppliers, and partners will arise as matters of chance rather than of managerial intentions.

The differing strategies of U.S. and Korean companies toward their respective governments' industry promotion programs illustrate these points. When U.S. government programs to promote the FPD industry hit stride, the international community of producers, materials makers, and equipment suppliers in Japan had already made substantial progress toward creating high-volume manufacturing processes for TFT LCDs. Korean companies took advantage of this progress by importing proven expertise and production equipment to create a basis for learning how to do high-volume TFT LCD manufacturing. During the same period, many U.S. companies implemented strategies to claim global market share under the umbrella of U.S. trade law and government FPD industry promotion schemes. Korean companies accepted government funding as well, but dragged their feet when invited to participate in government coordination efforts, competing aggressively with each other to establish themselves in the industry on their own terms. Companies that signed on to the most visible U.S. government-supported programs accepted limitations on joint development with suppliers and customers outside of the U.S. While in principle they were free to establish such relationships outside the U.S. once a U.S. partner had been identified, in fact few expended the resources to do so. Signing onto these programs also biased U.S.-based FPD producers toward co-developing and purchasing equipment from U.S.-based equipment makers who lacked high-volume production experience, rather than selecting the best equipment and partners on the international market. We will discuss details and implications of these different corporate strategies for FPD production in Korea and the U.S. later in the chapter.

## SPEED AND BUREAUCRACY

Competitive advantage in new, knowledge-driven industries depends on companies' capabilities to lead generational transitions that simultaneously increase product functionality and lower costs. Within the life cycle of each generation of TFT LCDs, productivity improved as yields increased, while market prices declined rapidly as companies added capacity. Transitions to new generations made price increases possible, because screen sizes increased and performance improved. Yet each generation's TFT LCDs arrived on the market at a lower introductory price than had their predecessors. The first companies to achieve commercial yields in each new generation enjoyed high profits due to the price increases and the productivity improvements associated with increased through-put of larger-sized substrates. Follower companies missed these high-profit windows, as their very arrival on the scene hastened price erosion. Managing these transitions rapidly depended on a company's abilities to build on the accumulated knowledge represented by its people, its relationships with equipment and materials suppliers, and its relationships with customers. In Japan and Korea, governments found they could exert little direct influence in these relationships and processes of change. The industry evolved consistently ahead of governmental conceptions.

Political processes do not run on Internet time. Decisions regarding government-funded programs generally must take into account the preferences of multiple funding sources, government agencies, and interested industry actors. The formal decision processes employed properly value equity, transparency, and fairness above efficiency and foresight, and they take a great deal of time. This need not pose a difficulty when governments' goals involve funding basic research or setting the broad rules of the economic game, such as the boundary between acceptable, constructive collaboration among companies, and anti-competitive tactics actionable under anti-trust law. Public policy programs aimed at creating specific opportunities, relationships, sectors or serving particular product or technology objectives in rapidly evolving industries, however, run the risk of misdirecting managerial attention toward fixed objectives that fade into the past more quickly than bureaucracies can identify and agree on them. The risk grows in proportion to the speed of industry evolution and the degree of focus in the policies and goals that government promotes. Meaningful participation in the critical turning points of a new industry's evolution requires physical presence and physical interaction on an ongoing basis in the laboratories, business meetings, and on the factory floors where knowledge creation takes place. Government officials cannot engage an industry in this way. Nor should corporate strategists who value the autonomy and efficiency of their companies' decision-making processes wish them to do so.

Joint U.S. and Korean government-funded efforts at matchmaking between U.S. equipment makers and Korean FPD producers offered a case in point. Contacts existed between the U.S. DoD and Korean government agencies including the Ministry of Science and Technology (MoST),[18] and between public-private display research consortia that existed in each country. The latter are known as the Electronic Display Industrial Research Association of Korea (EDIRAK), established in 1990, and the United States Display Consortium (USDC), established in 1993. Both consortia disbursed government funds, matched by recipient companies, for projects that served industry priorities as agreed by the organizations' governance groups. Governance consisted primarily of industry representatives, with minimal professional administrative and government representation. Each organization restricted membership until the late 1990s to companies at least 50-percent owned and controlled by citizens of its respective country.

Overtures for U.S./Korean cooperation within the USDC framework began at least as early as 1994, when the organization rebuffed the membership requests of LG, then known as Goldstar, and Samsung. Goldstar's general manager of LCD marketing and sales, K.Y. Lee, commented, "We have been investigating a possible partnership in the U.S. for several years. The problem is finding an American company to commit to production. They can't make a firm decision on their own, and are always waiting on government support before they move ahead."[19] Some U.S. government officials, for their parts, actively encouraged the USDC to find ways to cooperate with the Korean firms.[20] USDC officials remained cautious concerning the political flexibility afforded them, given congressional oversight of their mandate.

Two years later in October 1996, cooperation emerged in the form of a memorandum of understanding (MoU) between EDIRAK and the USDC and a visit of twenty-six member companies, along with DARPA officials, to Korea. The MoU envisioned funding joint projects that would partner U.S. equipment and materials suppliers with Korean producers. Implementation of specific projects was slowed, however, by complex negotiations to work around both organizations' practices regarding the funding of cooperation with non-members.

The torrid pace of industry evolution during this period produced an additional conundrum, as the circumstances of the Korean manufacturers changed radically. Between 1994 and 1997, when the first joint EDIRAK-USDC projects were awarded, both Samsung and LG successfully ramped up high-volume Gen 2 and Gen 3 TFT LCD lines. Hyundai had successfully ramped up Gen 2 production. As trailing movers facing stiff price competition, none of these companies were willing to risk the potential damage to their cost structures of working new, unproven equipment into critical stages of their high-volume production lines. Somewhat to the surprise of some American participants, EDIRAK did not have the power or mandate to persuade them to do so.[21] When the Asian financial crisis slowed fab investments in 1998, the projects remained in limbo. The board of one U.S. participant, MRS Technologies, a supplier of high-resolution steppers, compelled management to abandon its project, along with its commitment to the industry. USDC funding had permitted MRS to resume a stepper development project idled by a cutback in DARPA funding two years earlier.[22] This time, the clock ran out on top management as well, as respected industry leader Griff Resor stepped down from the company's presidency.[23]

### LEARNING TO KNOW WHAT YOU DON'T KNOW

In the early 1990s, prospective high-volume FPD industry entrants in the U.S. and Korea faced the same dilemma, but viewed it very differently. The companies setting up high-volume production in Japan—particularly Sharp, NEC, Toshiba, and IBM—had opened up a gap between themselves and companies producing elsewhere. Developments in Japan had established the clear promise of TFT LCDs as the first FPD technology to threaten CRTs in any application, as well as heralding a new multimedia era in applications for which CRTs could never have served.

Prospective Korean entrants viewed the breach primarily as a knowledge gap. They looked around the world, including to Japan and the U.S., for means to learn to bridge it. At Samsung and LG, decisions were taken to acquire state-of-the-art production equipment for pilot manufacturing lines, using these as platforms to ramp up to production on a scale consistent with plants located in Japan. Hyundai acquired a controlling interest in a U.S. company, ImageQuest Technologies, Inc., run by active matrix pioneer Scott Holmberg. ImageQuest's low-volume production line manufactured extremely high-performance TFT LCDs using mainly U.S.-built equipment and served as Hyundai's pilot facility. More than sixty Hyundai engineers trained there. When Hyundai built its first high-volume facility in Korea, however, the engineers brought it up to scale with the assistance of orders and advice from Toshiba, following production line models developed in Japan.[24]

Most prospective U.S. entrants viewed the gap in financial and market structure terms. They cited Japan's low cost of capital in the late 1980s and local availability of manufacturing equipment and materials as root sources of geographic concentration of production. Convinced of U.S. superiority in product technology, the companies looked to domestic semiconductor equipment makers for tools and materials to catch up in manufacturing, government and venture capitalists for financing for domestic fabs, and niche product markets, such as defense, avionics, and medical imaging, to absorb the output.

By the mid-1990s, neither venture capitalists nor the U.S. government were capable of generating the one-half-billion-dollar up-front financing and again as much to ramp up production that experts estimated would be necessary to put even one new company in the high-volume game. Risk-averse venture capitalists lacked the global perspective to accurately assess the threats and opportunities posed to U.S.-based production by the industry's geographic concentration in Japan. Government officials wanted to encourage U.S. production, but were constrained by domestic politics and international trade law from supporting such a concentrated investment in manufacturing. The General Agreement on Tariffs and Trade (GATT)[25] prohibited government subsidies for production of goods moving in international trade, but permitted government support of R&D. Consistent with the U.S. commitment to these norms, the government channeled R&D incentives to potential "next generation" equipment makers, companies that would test new equipment in production, and companies investigating TFT LCDs as well as potential leapfrog FPD technologies, such as FEDs. But niche markets would never absorb the scale of production needed to advance the manufacturing technology and drive costs down.[26] As a high-volume competitor said of the small-scale U.S.-located TFT LCD fabs that emerged, mainly with DoD support, "Mathematically, running at full capacity for an entire year, they can't break enough glass to learn to make a display."

In order to succeed, companies needed independent vision and means, along with a certain humility and openness concerning the globally interdependent nature of the knowledge-creation process in the industry. Companies needed to learn to know what they did not know,[27] and where to look for it. This paradox did not prove easy for many U.S. companies—large or small—to digest.

### Mapping, Catching, and Riding the FPD Industry's Evolutionary Spiral

The distinctive FPD business models that emerged in the U.S. and Korea corresponded to radically different conceptions of the role of knowledge accumulation in FPD industry advance, as well as the nature of knowledge-driven competition. In the early 1990s, companies in both countries demonstrated capabilities to advance the state-of-the-art in FPD product technology. In the U.S., for example, the Xerox FPD spinoff dpiX demonstrated an extremely high resolution, 6.3-million pixel, 13-inch TFT LCD with the capability to perfectly emulate print on paper.[28] In Korea, Samsung innovated in TFT LCD brightness by increasing the aperture ratio for its displays, and credibly sought to lead the industry in increasing the size of TFT LCDs that its lines could fabricate from single substrates.[29]

As trailing movers in large-format TFT LCD production, companies from both countries stood in position to acquire critical embodied process knowledge in the form of Generation 2 production equipment and materials. Acquiring state-of-the-art equipment and materials would not in itself, however, suffice to establish companies in the high-volume arena. The companies would need to acquire unembodied knowledge along with the equipment, some portion written and available in the form of manuals and licenses, some not. Some, but not all, of the unwritten knowledge that producers in Japan had accumulated by bringing up production lines could be acquired through the services of consultants. Some part remained tacit or team-embedded and could never be transferred except by acquiring the permanent services of engineers and operators who had awakened or worked on original TFT LCD lines. Such individuals were available, but in very limited supply.

The high speed of industry evolution assured that the proportion of verbally transmitted and tacit knowledge needed for efficient operation of production lines remained significant through each generation's life cycle. Critical bodies of knowledge feeding into the design of new equipment generations, therefore, also remained largely verbal, tacit, or both, as did much indispensable knowledge for speedily managing generational transitions and bringing up new lines. In order to position themselves to bring up future generation production lines in a timely fashion, companies needed to establish their own, proprietary experience bases on existing generation, high-volume equipment. This process started with providing individuals and work teams with the experience of integrating individual pieces of equipment into functioning production lines and bringing the lines up to competitive yields. These individuals' knowledge, the knowledge shared within their teams of coworkers, the relationships they developed with equipment and materials suppliers, and the shared knowledge in these relationships, in turn, would inform the new equipment design and integration process for subsequent generations. This interaction among producers, equipment suppliers, and materials manufacturers grew in importance for FPD producers' performance as competition tightened and generation changes speeded up. Producers increasingly sought to create or extend competitive advantages in ways that demanded equipment and materials customization.

Successful FPD producers did not enter an industry so much as they entered a knowledge stream. The difficulty for prospective U.S.- and Korean-based producers arose in trying to figure out where to jump into the stream. In the mid-1990s, analysts typically calculated the cost of industry entry as the sum of prototyping, characterizing a manufacturing process, building a third or third-plus generation TFT LCD fab, and bringing it up to commercial yield. Alternatively, some critics suggested that inexperienced entrants building third-generation fabs were really building past-generation fabs. They calculated the cost of entry as a plunge into the infinite.

### Back to the Future

Both views were too narrow. The knowledge-creation processes that drove TFT LCD technology ahead demanded a multi-generational commitment that required entrants to look backward and forward at the same time. Forward-looking strategies suggested that companies should build next-generation fabs, timed to achieve commercial yields before a subsequent generational transition shifted the industry's cost structure downward. In fact, companies also needed investments in current generation fabs to prepare themselves for profitable participation in the next generation. Next-generation fabs built on equipment and experience gathered in past and current generations. This is why, in Figure 7.4, we depicted the role of knowledge in FPD industry advance as a spiral rather than a straight line.

The niche strategies of most U.S. firms aimed to defer or sidestep these realities by differentiating products for markets, such as defense and medical imaging, that managers hoped would absorb prices far above the average for notebook displays. Company strategies called for establishing small fabs to serve these high-end markets, generating fresh technical foundations as well as cash flows to help support an eventual, autonomous leap to high-volume production.[30] Their fabs used a much higher proportion of U.S.-origin production equipment than did the manufacturing lines established in Japan. Much of this equipment represented manufacturers' initial forays into the business, had no history of use in prior generations, and proved difficult to integrate and raise to commercial yields.[31] Although the fabs were less automated and more labor-intensive than the high-volume facilities in Asia, bringing up and operating them did not prepare a sufficient cadre of engineers and operators to guide ramp-ups of future generations.

The Korean entrants set strategic objectives to profitably seize both differentiation and cost leadership, by establishing primacy or at least close followership in the transitions to Gen 3 and subsequently Gen 3-plus high-volume production technology. They also pushed process technology forward through other productivity enhancements, including increased array testing, inspection, and cleanroom particle control.[32] Running state-of-the-art Gen 1 and Gen 2 lines at pilot quantities, the companies began to accumulate experience to selectively enter equipment and materials manufacturing as well as high-volume Generation 3 production. Samsung, for example, achieved commercial yields on its first high-volume line, a Generation 2, in July 1995,[33] at approximately the same time as Sharp and DTI were starting up their Gen 3 lines. The company started up the industry's next Gen 3 line in October 1996[34] and broke ground for a Gen 3-plus line to handle 600 by 720 mm substrates in January 1997.[35] During the same period, the company developed independent materials and equipment capabilities in color filters, driver chips, substrates (in its Samsung-Corning joint venture), and tab bonding, with plans to enter steppers.[36]

In our terminology, the U.S. companies adopted strategies narrowly based on product positioning, hoping to build down to high-volume applications from the high-cost, high-performance end of the FPD market. The Korean companies adopted strategies that reflected the high-speed knowledge-driven competition that already existed in the global industry, in which product and process technology advanced in tandem. These alternative perspectives on competition placed the U.S.-based producers on a collision course with the Korean producers, as well as with high-volume TFT LCD producers elsewhere. As producers continued to lower costs and learn to manufacture increasingly advanced TFT LCDs at high volume in Asia, the niche markets dear to U.S.-based producers came increasingly under assault.

In the end, even the U.S. military comparison shopped.

### The Wrong Butterfly

*Instrumentation displays for the Space Shuttle, the most sophisticated and intrepid human transport system yet devised . . . large, high-resolution, way-better-than-CRT, direct-view video panels for the command posts, situation rooms, and airborne observation centers of the most powerful military in the world . . . extremely rugged screens for tank-borne weapons targeting systems that can locate adversaries in the dark, or on the other side of the horizon . . . avionics displays that convey clear images in the brilliant sunshine of supersonic jet fighter cockpits . . .*

In the mid-1990s, U.S. companies produced made-in-U.S.A. high-performance TFT LCDs to meet each of these needs. Every one of these advanced applications supported critical, split-second decision-making with lives hanging in the balance. Yet by 2000, despite distinguished technologies and a successful history of seeking U.S. government contracts and other support, management in every one of these companies had chosen to leave the FPD industry or shut down key defense supply lines. These companies played central roles in joint business-government efforts to establish FPD production within the U.S. Did their demise represent the closing drama in a dream that turned into a nightmare, as one industry newspaper suggested,[37] or one of many informative outcomes in an experiment that met an important U.S. national need? If the latter, what lessons ought to be taken from the exercise?

If anyone doubted the vulnerability of U.S. national security interests to FPD supply interruptions in the mid-1990s, they had only to recall the consequences of a fire at Optical Imaging Systems' (OIS) new Northville, Michigan, plant in March 1995. The fire was small, confined to one piece of equipment and a duct. But the company had just finished installing $100 million worth of brand new cleanrooms and equipment financed with a $48 million DARPA grant matched by parent Guardian Industries. The thin film of ash that covered every surface rendered the new facility unusable. During the nine months that OIS employees spent cleaning up the plant with toothbrushes, according to the company, "every major U.S. weapons program was delayed."[38] An image arose from the incident that haunted discussions of FPDs and military procurement: a flightless fleet of F-16 jet fighters mothballed somewhere in a desert with cavities in their instrument panels where FPDs were supposed to go.

The new OIS facility served as one of two TFT LCD manufacturing test beds in the bundle of existing FPD industry promotion programs the Clinton administration combined in 1994 to create the National Flat Panel Display Initiative (NFPDI). The initiative established the second testbed as a consortium of three companies, Xerox (dpiX), AT&T, and Standish Industries, Inc. Funding for the USDC, which was founded in 1993, was also pulled into the NFPDI, as were other DARPA FPD projects started under the High Definition Systems (HDS) Program. The NFPDI showed promise of improving coordination and effectiveness in an area of critical need and considerable, but fragmented government resource commitment. The combined programs had disbursed grants of around $500 million since fiscal year 1991.

The program rationale was to provide the U.S. military with early, assured, and affordable access to leading-edge FPD products and technologies. This rationale gained urgency from Japanese suppliers' well-documented reluctance to supply U.S. military requirements, as well as questions regarding the legality of these companies' doing so under Japan's post-World War II constitution. The task force report that proposed the program also raised concerns that dependence on foreign sources for critical FPD components could stifle U.S. companies' competitiveness in civilian electronics markets.[39]

The initiative aimed to coordinate government efforts to encourage equipment and materials R&D, develop process technology, and to continue to push product technology forward. Companies "demonstrating firm commitments to volume manufacturing" were eligible for "focused R&D incentives" to encourage development of next-generation product and process technologies.[40] Consistent with prior policy, the initiative's designers enshrined technology neutrality as a funding criterion. Projects for TFT LCD, EL, FED, Plasma, and other FPD approaches all received support. The initiative was not in the business of picking winners and losers among either companies or technologies. Furthermore, the initiative's designers rejected the idea of funding a dedicated supplier to meet military needs. Affordability and the need for continuing innovation at the FPD technology frontier required a U.S.-based industry comprising firms that competed in global, high-volume consumer markets. The designers did not believe that government-led industrial strategy could summon such a high-volume industry into existence. Rather, they argued, the government should offer carefully calibrated incentives that might, at the margin, reduce U.S. managers' perceptions of the risks associated with participating in the high-volume industry, so that they would take action themselves.[41]

One senior manager of a company that had benefited from several U.S. government R&D contracts used a metaphor from non-linear systems theory to describe the intended subtlety of the government's interventions. As the physicist and best-selling science writer M. Mitchell Waldrop famously explained, "the flap of a butterfly's wings in Texas could change the course of a hurricane in Haiti a week later."[42] The point was, the manager went on to say, "we would not exist without DARPA."[43] This conception of the NFPDI responded in an arresting way to criticisms typically leveled at heavy-handed government industrial strategies. But the second part of Waldrop's analogy also needed to be taken into account: ". . . a flap of the butterfly's wings a millimeter to the left might have deflected the hurricane in a totally different direction."

In fact, the NFPDI did not unfold as an instance of heavy-handed government strategy superseding corporate strategy, but rather as one of many episodes in a pattern of business-government strategic interaction that began earlier in the decade. Several companies took the lead in politicizing U.S. FPD industry development, beginning with the 1990 formation of an industry association, the Advanced Display Manufacturers of America (ADMA). The seven founders included OIS, Planar, Plasmaco, Photonics Technologies, Magnascreen, Cherry Corporation, and Electroplasma, all companies that had received DARPA contracts. On July 17, 1990, the ADMA filed an anti-dumping petition[44] with the U.S. International Trade Commission (ITC), charging thirteen Japanese companies, including Sharp, Toshiba, Hosiden, and Hitachi, with predatory pricing of TFT LCDs. The commission authorized an investigation of the Japanese companies' production costs. Taking into account low production yields, the investigators concluded, fair market value for some of the companies' products exceeded the FPD prices on offer in the U.S. market.[45] Steep anti-dumping duties were authorized for several Japanese companies' TFT LCD products on August

15, 1991, at just about the same time Sharp and DTI were bringing up their first Generation 1 lines. But in November 1992, OIS, which had been recently purchased by Guardian and was the only U.S. domestic TFT LCD producer, requested that the duties be removed. On June 21, 1993, the U.S. Department of Commerce complied.

Despite its apparently innocuous conclusion, the anti-dumping petition permanently affected the course of FPD industry development within the United States. Notebook producers, faced with the prospect of paying tariff-laden prices for the most advanced displays, immediately moved their assembly operations offshore. U.S. Customs officials had ruled that the duties could not be levied on screens already incorporated into assembled goods.[46] The duties also placed an artificial price floor under TFT LCDs at a time when companies were struggling to move enough panels to drive production learning processes. Companies struggling with ramp-ups of new fabs in Japan found they could charge close to the tariff-burdened price for displays selling there and to notebook assemblers producing in third markets. "This was an unexpected windfall," a respected former FPD market analyst later suggested. "The TFT manufacturers were able to put together quite a war chest, which allowed them to expand capacity more rapidly than expected."[47]

The petition also validated a bias in many U.S. companies toward framing the industry-knowledge race in terms of international rivalry among countries rather than global competition among firms. Many continued to look to government for the resources to compete. The widespread impression among U.S. industry participants held that the government needed to step up its involvement in the industry to counter Japanese government investments. In fact, any such investments were minimal[48] and had directed companies' resources to a technological dead end that was subsequently abandoned.[49] The U.S. public television documentary series *Frontline* offered a one-sided assessment of the anti-dumping case and its aftermath, asserting that Japanese government support had played an important role in establishing the industry in Japan.[50] One defense industry journal reported with expansive inaccuracy in May 1993 that, "the Japanese cornered LCD manufacturing capability by government investment of almost $4 billion."[51] None of these reports reflected firsthand experience of industry circumstances in Japan. But in retrospect, they evoke the atmosphere of national urgency in which AT&T, Xerox, Standish,

OIS, and the members of the ADMA entered negotiations with DARPA in 1993 to jointly fund an R&D consortium to help jump-start the industry in the United States.[52]

The discussions concluded with the establishment of the USDC on July 20, 1993, as a non-profit, public/private consortium with a primary mission of supporting the development of an FPD manufacturing infrastructure in the U.S.[53] During its first six years of existence, the organization consisted of FPD producers, users, and equipment and materials suppliers with at least 50 percent U.S. ownership. The group based its structure on that of SEMATECH, another public/private consortium formed by DARPA and U.S. semiconductor producers and equipment makers in August 1987. According to one of several press releases issued to announce the consortium, however, important differences existed between the two programs. Unlike SEMATECH, the USDC would not establish an R&D and pilot manufacturing facility in which to test new equipment and materials.[54] This approach had not worked well for SEMATECH, because semiconductor manufacturers that were engaged in their own equipment development programs were reluctant to share a common factory floor.[55] USDC development programs called for member manufacturers to test new equipment and materials in their own commercial fabs.

The absence from the membership rolls of high-volume manufacturers who could fulfill this role,[56] however, undermined the USDC's mission to "build the U.S. infrastructure required to support a world-class, U.S.-based manufacturing capability." As the centerpiece of its programs, the consortium identified U.S. industry development needs and invited proposals from members for projects to meet these objectives. Development teams consisted of equipment and materials suppliers working with an FPD producer that would serve as project coordinator and beta site. The USDC provided grants to defray project costs out of its DARPA funding, which the winning bidders matched at equal or greater value.[57] But the USDC membership framework did not provide members with development partners who could qualify and integrate their equipment and materials innovations in the global, high-volume manufacturing context. No high-volume TFT LCD manufacturers existed on U.S. soil. Even if one had existed under foreign ownership, USDC practice[58] would have proscribed contracting with it.

The issues that interposed between many U.S. equipment and materials manufacturers and high-volume producer/development partners reflected managerial

mindsets as well as consortium policy and practice. Industry officials with influence over the consortium's project selection process did not believe that interdependence among equipment, materials, operators, and R&D scientists differed
in any meaningful way between low- and high-volume production lines. Some
did not regard the matter of line integration as important at all in designing new
equipment, asserting that new pieces of equipment could, in principle, be qualified for high-volume production with data generated by "running them by themselves for a few days in a room."[59] But the question was not one of principle, but
rather one of practice. In practice, Generation 1 high-volume production lines
were already running at high yields by the time the USDC's programs were
established, and their operations had for some time been contributing vital
knowledge to the design of Generation 2. Competing with existing equipment
and materials makers would require companies to demonstrate a capability to integrate into existing production line systems, while making a clear contribution
to both product features and yield enhancement. Participants in USDC development programs might have greatly benefited from opportunities to integrate new
tools and materials into lines that incorporated process solutions reflecting the international state-of-the-art. This would require beta-siting in a production context with equipment and materials of diverse international origins.

U.S.-based producers, however, gave priority to U.S.-origin equipment when
they established their fabs. At OIS, executives apologized for the few Japanese-
origin tools on the production line.[60] Executives at Hyundai's ImageQuest affiliate in Fremont, California, expressed pride in creating a production line and
process using equipment originating almost entirely in the U.S. "We're more
American than the USDC," president Scott Holmberg commented during a
fab tour, noting as well that USDC ownership rules at the time precluded
ImageQuest from membership.[61] USDC members wishing to qualify their project outcomes in a state-of-the-art production context needed their own international contacts and resources to do so. Photon Dynamics, whose project ranks as
the USDC's most significant global success, was already working closely with
Japanese and Korean customers as well as investors when it accepted the USDC's
first contract for a TFT LCD visual inspection system.[62] Few other members
enjoyed similar advantages.

Given that global interaction among equipment suppliers, materials suppliers,
and producers has driven the evolution of product and process knowledge across
TFT LCD generations, any of three alternative strategies might have served to

address these limitations. The first two would have required resources beyond those available to the USDC, but might nonetheless have proven possible within the larger context of DARPA FPD programs. The third possibility would have involved a change entirely within the USDC board's discretion. In taking the first alternative, the USDC would have reverted at least partially to the original SEMATECH model. The consortium might have set up an R&D line using state-of-the-art equipment sourced in international markets, providing members with opportunities to integrate and qualify their offerings. Alternatively, the consortium might have supported the U.S. establishment of a high-volume commercial facility, using proven technology. Discussions between DARPA and IBM to license DTI's process technology to a U.S. consortium of companies, however, foundered on issues variously reported as timing, concerns about violating program principles of technology neutrality, and the political difficulty of involving U.S. taxpayers' funds in a program that might pay royalties to a foreign entity.[63]

Ultimately, the USDC chose the third option. In 1999, it altered its charter to admit foreign-owned companies to associate membership.[64] By then, however, there existed little alternative. OIS had announced its shutdown on September 18, 1998.[65] On March 12, 1999, dpiX management announced to staffers that the organization would begin a sixty-day wind-down period, preparatory to closing its doors.[66] Military orders had proven insufficient to keep these two TFT LCD makers afloat, particularly considering that the Pentagon was using comparison costs on products available from high-volume Asian producers to set prices.

Planar Systems, the world's leading EL display manufacturer, became the only TFT LCD producer remaining on U.S. shores. It had been ruggedizing panels for military applications and led an acquisition of its cell supplier, dpiX, in partnership with unnamed consortium partners.[67] In January 2000, however, Planar's top management notified the DoD that the company would cease production unless it could raise its prices to cover costs. The DoD had been paying $12,000 per panel. Planar claimed costs of $43,000 per piece. In September, after months of unsuccessful haggling, the company announced that it would drop the business.[68]

In a bizarre twist on a story that DoD officials had used repeatedly to dramatize the need for an NFPDI, a company had declined to comply with U.S. military requests. But this time, the company was American, not Japanese.

## SWIMMING IN THE KNOWLEDGE STREAM

Perhaps it was jet lag or the gray skies of the oncoming Korean winter, but on December 5, 1996, the prospects for a productive visit to Samsung's TFT LCD facility did not seem promising. Sequestered and awaiting clearance in a small reception building on the fringe of Samsung's city-scale Kiheung complex, scores of vendors, subcontractors, customers, and other visitors milled around and smoked. Far more slowly than the stream of new arrivals accumulated, Samsung hosts wearing one of several different brightly colored blazers appeared to escort them into the complex. Jun H. Souk, executive director of AMLCD R&D, arrived wearing a pewter warm-up jacket emblazoned with the company logo, offering a formal, but distracted welcome. The colleague who had agreed to join him in meeting three visitors from the U.S. had been unexpectedly called away to another meeting in Europe. Time was limited. Top management had just requested Souk to move the profitability target for Samsung's TFT LCD business far forward on the calendar into the coming twelve months. The executive director would need to speed up every aspect of the operation to accommodate. Meetings with business researchers diverted attention from reviewing daily output and yield figures, which were very promising, but not has high as Souk would have preferred.

Souk missed his colleagues at IBM's Thomas Watson Laboratories in Yorktown Heights, New York. But he wanted hands-on involvement in high-volume manufacturing, not just product and process research, and he wanted to run a TFT LCD business. Like Kye Hwan Oh, Hyundai's executive vice president of display devices and electronic components, Jun Souk is a naturalized U.S. citizen. After a Ph.D. in solid state physics from Ohio State University and ten years at IBM, he returned to Korea. The new job was stressful, but the opportunity to participate directly in establishing a high-volume FPD business was too exciting to pass up.

The visitors had recently paid a call at Yorktown Heights. "You know Steve Depp?" Souk's manner modulated from formal to expansive. He relaxed. "Well, now I am getting interested!"[69]

The differences between the stories of FPD industry emergence in the United States and Korea turn largely on relationships. Managers in many, although not all, U.S. companies chose to turn inward, toward each other, and toward government to locate themselves in the processes that were building the new industry. Korean companies turned outward. Government engagement took a similar external orientation, tempered by concern that the successful ramp-up of the industry in Korea depended too heavily on Japanese equipment, materials, and technical input. Government officials viewed international diversification of these relationships as an important priority and hoped that corporate managers would see things the same way. "We see many proposals," the Ministry of Science and Technology's Jae-Choon Lim commented. "When a project contains international cooperation, we accept 100 percent."[70]

Like the successful U.S. and Japanese FPD companies, Korean producers did not wait for government action before opening up their international channels. Although Korean government guidance suggested an alliance to establish TFT LCD production in Korea, management at Samsung, LG, and Hyundai chose to enter the industry independently and compete with each other. Distinctive approaches to international collaboration provided sources of competitive advantage for all three entrants and helped two of them—Samsung and LG—win the two leading global market share positions by 2000. The same independence complicated efforts by U.S. and Korean public/private consortia to broker further collaboration. "Korean companies are interested in independent relationships with U.S. companies," a USDC board member commented in 1996. "We have to keep reminding them that we are a consortium."[71]

These independent international relationships took three forms: technical cooperation, strategic alliances, and long-term contracts. Some relationships contained elements of all three. Each type of relationship actively contributed to the companies' capabilities to address the basic challenges of knowledge-based competition: *learning, continuity,* and *speed.*

*Technical cooperation* included equipment and materials supplier relationships, customer relationships, and R&D partnerships, including licensing. Technical cooperation relationships helped companies establish a knowledge base in current-generation production technology, move rapidly into production, and create a foundation for continuous learning in ramping up successive new-generation facilities. The Korean companies' positions as close followers to companies that had established high-volume production in Japan offered both advantages and challenges. They could and did purchase equipment, process recipes, and extensive consulting services from the successful producers, equipment manufacturers, and materials makers. As a consequence, at Samsung and LG, Generation 2 installations came on-line and reached commercial yields relatively quickly—but not quickly enough to take advantage of the profits available to first movers.

Samsung and LG gained critical knowledge advantages by ramping up their Gen 2 lines, however, even in the face of price declines. Already committed to Generation 3 investments in the range of $600–800 million, both companies needed to leverage the knowledge gains from Generation 2, particularly experienced operators, to move rapidly forward. Samsung entered Generation 3 in late 1996, reaching commercial yields in early 1997, hot on the heels of DTI and Sharp. LG followed with its Generation 3 line in the second half of the year, but running a slightly larger substrate that offered cost economies while optimized for slightly larger displays.

Technical cooperation relationships as well as *equity-based strategic alliances* also helped the companies to cut costs in face of continuous price declines and to differentiate their products. Samsung's alliance with Corning, Samsung-Corning Precision Glass Co., placed it alongside the leading substrate supplier in the forefront of glass innovation. Samsung-Corning opened its first fusion glass plant in Korea in 1995.[72] The relationship contributed to increased efficiency and helped Samsung approach generational transitions with confidence and foresight. In 1995, Samsung entered into a cross-licensing agreement with the Japanese firm, Fujitsu, a fellow late TFT LCD entrant. Fujitsu provided its wide viewing-angle technology in exchange for Samsung's high-aperture ratio, brightness-enhancing technology.[73]

LG management regarded technical cooperation as an even more central element in strategy, in part as a means of compensating for the company's size difference with Samsung and Hyundai. "Our philosophy is not to try to do everything for ourselves," said Choon-Rae Lee, managing director of LG's LCD Division. "We will work with anyone who can add a cost or differentiation advantage."[74] Management also set a goal to excel in particle control and yield enhancement. At least two technical cooperation agreements significantly contributed. In 1994, LG entered a $30 million joint venture with Alps Electric, a Japanese components firm, to develop ultra-clean manufacturing technology at Alps Central Laboratory in Japan. LG implemented the technology for the first time on its Generation 3 line at Kumi.[75] Its work with Photon Dynamics on TFT array test equipment proved crucial to meeting LG's zero-defect objective,[76] and helped the company gain a five-year, $1 billion contract to supply 12.1-inch displays to Compaq, despite having only one year of volume production experience.[77]

*Long-term contracts* as well as equity-based alliances with customers played an important role in sustaining continuity. Only Hyundai delayed ramping up its Generation 2 line, which it had installed by the end of 1995, hoping for stabilization in 10.4-inch prices.[78] Technical cooperation tied to a long-term sales agreement with Toshiba helped the company to overcome subsequent delays in achieving commercial yields[79] and to reduce further delays in moving to Generation 3. Hyundai's transition to Generation 3-plus, like that of all of the Korean producers, was complicated by external events of global significance.

Financial crisis gripped Asia in the late 1990s, placing the Korean TFT LCD producers' ambitious expansion plans at the mercy of an investment capital crunch. Long-term contracts assumed increasingly vital roles in helping to continue next-generation investments, while at the same time assuring notebook computer companies of an increasing supply of the most advanced display components to sustain their growing businesses. In November 1999, Hyundai concluded contracts with four notebook manufacturers—including IBM, Compaq, and Gateway—for five years' sales of $8 billion.[80] In March 2000, Hyundai announced that it hoped to start up a next generation fab at Ichon, raising the company's planned production capacity to 1.5 million TFT LCDs annually.[81]

In July 1999, Apple Computer revealed plans to invest $100 million in Samsung in order to speed the construction of new production capacity for TFT LCDs.[82] In October 1999, Samsung signed a five-year contract worth $8.5 billion to supply TFT LCD displays to Dell Computer Corporation.[83] Having doubled capacity in 1999, Samsung was on track to open the world's first fab to utilize 730 by 930 mm substrates.[84] Industry sources differed on what number to designate the new generation. One called it "Generation 3.7,"[85] others 3.5-plus. Samsung preferred "Generation 4." Many industry participants still waited for a fabled one-meter square substrate to bear that designation.

The wild duck spirit was alive in Korea. Management decisions to expand production and continue TFT LCD generational progressions in the face of the Asian financial crisis surprised industry observers. But it was consistent with the high-volume industry's early history of countermanding financial intuition in decisions to move ahead. Those early decisions were made, in the words of a senior LG executive, in a "not logical thinking way."[86] Similar vision thrust Samsung and LG well ahead of the more cautious producers in Japan, as well as the paralysis in the United States, and created two very profitable businesses.

LG's TFT LCD business was so profitable, in fact, that management struck a defiant pose when government's crisis plans for restructuring Korean industry demanded the combination of LG Semiconductor with Hyundai's semiconductor business. Unhappy about any such plan, management made it clear that the LCD Division's assets, with a book value of about $1 billion, were not on the table.[87]

International markets ratified management's decision with the May 1999 announcement that Royal Philips Electronics of the Netherlands would acquire 50 percent of LG's LCD unit in exchange for an investment of $1.6 billion in the joint venture. LG.Philips LCD was established in July 1999 and officially began operations in September 1999. Bon-Joon Koo, former CEO of LG Semicon, was named CEO of the new company.[88]

In 1999, Samsung's global FPD market share, ranked first, stood at 18.8 percent. LG.Philips' share, ranked second, stood at 16.2 percent. In a few quick moves, Korean companies, staffed by many U.S.-educated engineers and managers, broke Japan-based sources' short-lived, near-monopoly over high-volume, large-format color TFT LCD production. Furthermore, the Korean companies did it with the cooperation of equipment makers, materials producers, and TFT LCD producers centered in Japan. Philips established a European presence in high volume even more rapidly, seizing opportunities for TFT LCD production partnership that every U.S. company except IBM had neglected for years. Managers of the Japanese companies, Korean companies, and the American and European companies that partnered with them realized that quick moves were what it would take to establish their roles in a new, knowledge-driven industry. A few quick moves offered the only chances they would get to be a part of it. It was, as IBM's Hirano said, "a gamble business." As an earlier generation of high-tech entrepreneurs might have said, they let the chips fall where they may.

# 9

# Knowledge and Transcendence

*TFT has moved very fast. One would be*
*lucky to catch up.*

HSING C. TUAN[1]

**W**e set out to write this book with two objectives in mind. We wanted to show how the knowledge basis of new industries requires a new way of thinking about competition. And we wanted to show how thinking in this new way could enrich our understanding of the emergence of an important new industry from which we can gain insights for the future.

Early events in an industry's history exert powerful influences on the future path of its evolution. The geographic location of pioneering companies that commercialize a new technology, the nature of that technology, early design wins, and market successes all seem to conspire to lock in an industry structure, locational advantage, and a permanent cast of competitors that can only change through attrition.

Economic historians call it "path dependence."

Path dependence makes a poor assumption for strategy in new industry creation. It relieves management of too many important choices too soon. One perspective might suggest that the ideal strategy sets the path by establishing a dominant design. In knowledge-driven competition, however, multiple designs overlap and rapidly succeed one another. Dominance can prove fleeting and potentially illusory for managers who assume their companies have achieved it. The most successful companies build learning capabilities that enable them to lead rapid change, transcending old advantages in the process of building new ones.

The capital intensive nature of FPD manufacturing and the apparent concentration in Japan of the significant capital investments needed to found the industry profoundly misled many strategists about the possibility of entry and the necessary success factors. In fact, investments in physical assets and Japanese locations, while important, did not constitute the critical determining factors for success. After all, the early high-volume production lines proved virtually inoperable when they were first installed.

Investments in knowledge creation, including product innovation, manufacturing process innovation, and manufacturing line integration, made the difference between success and failure. Some of the new knowledge that accumulated could be written down and became embodied in manufacturing equipment. This knowledge was foundational to industry progress but pertained to the past. Producers' most critical assets proved to be people, working as individuals and in teams, whose personal knowledge was essential to companies' capabilities to migrate rapidly from early generations of product and manufacturing equipment to future generations.

## LEADING RAPID MIGRATION

Capabilities to migrate rapidly across technology generations play a critical role in managing new industry creation because dominant product design and manufacturing process paradigms both remain unsettled. New generations appear in response to customers' requirements to increase product functionalities. But expanding the market for a new industry's products requires continually falling prices. Innovating companies cannot rely on product uniqueness to provide higher-than-average returns unless they can cut costs at the same time as they improve their products. Success requires them to couple product innovation with process innovations that increase manufacturing efficiencies. Late movers in generational transitions face extreme margin pressures because their entry increases supply and drives prices down at the same time as first movers in follow-on generations arrive with more desirable products and fundamentally lower cost structures. Consequently, only early movers enjoy much chance of profit, and sustaining profits depends on arriving early again at the next generational transition.

The ability to manage such rapid, continuous change depends critically on experienced people who carry a significant proportion of prior generation knowledge forward as part of the foundation for new knowledge accumulation. The importance of individuals and teams grows as the pace of industry evolution from generation to generation increases. This occurs because during periods of generational transition, personal and unwritten knowledge accumulates more rapidly than knowledge that is written or embodied in fixed assets such as capital equipment and materials.

## KNOWLEDGE AND FOLLOWERSHIP

Few companies enter emerging industries as founding members. Despite holding follower status when they start out, however, some companies play critical roles in building new industries and end up as leaders. Good followership counts for a lot in establishing a viable business in the knowledge-driven competition that defines new industry creation. Few companies can hope to outflank a technology leader by leaping a generation ahead. Yet many companies that considered entering the FPD industry after a generation or two of high-volume production technology had passed fell victim to the self-defeating notion that they had no other choice.

Stepping forward into ongoing, knowledge-driven competition begins by taking a step back, recognizing that the point of entry is not a teacher's position, but that of a student. Follower companies can often take advantage of equipment, materials, licenses, process recipes, and consulting services that encompass important elements of the knowledge created by predecessors who may have started from nothing. Creating the vital resources to succeed in a knowledge-driven industry, however, does not begin with purchasing state-of-the-art technology, but rather with creating a basis in people for learning how to use it. Often this means entering the industry with current generation technology, achieving commercial yields, and running at efficient scale to build up the knowledge foundations necessary to seize a leadership position as the next generation emerges. Substandard returns or losses that come with late entry in current generation technology amount to tuition, reimbursable through timely entry to the next. The leading large-format TFT LCD market shares held by Samsung and LG.Philips in 1999 and 2000 stand as testimony to the rewards of good followership.

**PATH INTERDEPENDENCE**

Between 1995 and 2000, companies from around the world entered into a learning competition that inevitably eroded Japan's FPD production dominance. The companies that built the Japanese production base also entered into a competition to profitably share the knowledge they had accumulated. In late 2000, Taiwan-based production sites were expected to capture 16 percent of the global market for TFT LCDs by the end of the year and to capture another ten points by the end of 2001.[2] Japan and Korea-based production sites were running neck and neck to divide the remainder of the market. The first AMLCD fab had been established in China.[3]

The world had seen such patterns of production and technology diffusion from Japan within Asia before, in DRAMS, for example.[4] The speed of the shift, however, was unprecedented. Some competitors, particularly in Korea, deprecated the technological capabilities of the Taiwan-based producers, explaining their emergence as a case of bottom-feeding on low-end, price-driven, mature market segments.[5] Even as such dismissive comments filtered through the media, however, some producers in Taiwan were leveraging their positions in small FPDs for such products as mobile telephones and pocket TVs to prepare for a broad-based market assault based on technological advances in products as well as process improvements.

In fact, the Taiwanese companies' entry strategies had much in common with the strategies Korea-based producers had undertaken in the early 1990s, with a principal exception of greater openness to direct intercompany collaboration. As such, they reflected the continuing vitality of knowledge creation as the industry's driving competitive force. Starting with proven Gen 1 and Gen 2 production equipment, materials, and processes, the new producers built a cadre of knowledgeable researchers, engineers, and operators capable of bringing up Gen 3-plus lines. Management recognized the essential relationship of speed and timing to profitability in ramping up current generation lines to high volume, and they formed learning alliances with Japan-based producers. Chung Hwa Picture Tubes (CPT), for example, established relationships with ADI, itself a joint FPD production venture between Asahi Glass and Mitsubishi. CPT accelerated the ramp-up phase on its initial Gen 3 line by sending operators to work for a time at an ADI fab in Japan. ADI operators also traveled to CPT for the startup of its line, which reached commercial yields in the second quarter of 1999.[6] Unipac, which had started up the first high-volume TFT LCD fab in Taiwan in 1993, aligned with Matsushita to ramp up a Gen 3.5 line to high volume by 1999's fourth quarter. Acer worked with IBM to deploy DTI process technology, Hanstar worked with Toshiba, and Chi-Mei set itself up as a foundry for Fujitsu.[7] Quanta Computer, Inc., which supplied notebook computers for Dell and other branded manufacturers, entered a joint venture with Sharp to build a fab, which was expected to begin ramping up in the first quarter of 2001.[8] Repatriated Chinese-Americans formed a backbone of the management and engineering staffs for many of these ventures.

## BEYOND JAPAN: BUILDING THE GLOBAL, KNOWLEDGE-DRIVEN INDUSTRY

Success in the high-volume FPD industry has depended on access and openness to global knowledge-creation processes, factors that have not changed in a decade of industry evolution. As FPD technology continues to evolve and find new applications, active engagement in new centers of knowledge creation will remain a necessary component of strategy for the most successful existing companies as well as new industry entrants.

When Japan-based producers collaborated in cannibalizing their collective dominant position in TFT LCDs and FPDs generally, they affirmed a central tenet of knowledge-driven competition. They sustained the competitive advantages they had created by sharing and transcending them. These strategies helped in at least three ways to reignite and increase the pace of industry evolution in the wake of the Asian Financial Crisis, benefiting the companies as well as the health and competitiveness of the industry as a whole.

First, by licensing their processes, the pioneering manufacturers extended their windows of profitability for the Gen 3 production phase without investing additional scarce capital. At the same time, producers were able to apply the lessons of Generation 3 to move on rapidly to subsequent manufacturing generations.

Second, cross-national, intercompany collaboration to expand the industry helped to restore the health of the manufacturing equipment sector, which had been severely damaged by cancelled orders and delayed fab construction schedules. Rapid generational transitions depended in part on equipment and materials makers' capabilities to quickly generalize industry knowledge and spread the associated development costs for new generation equipment across a large number of customers. The industry needed to significantly increase in size to support an efficient scale of operations for makers of equipment such as steppers and CVD. In the wake of the Asian financial crisis, sustained industry expansion depended, in part, on attracting new players to broaden the pool of available investment capital. Expansion of the industry into Taiwan opened the market for additional sales of Gen 3 production equipment. Slower industry expansion might have thrown more TFT LCD producers on their own resources to vertically integrate into loss-making equipment and materials ventures. Such a trend ran the risk of reducing industry knowledge flows, slowing knowledge accumulation, and increasing costs at a time when next-generation fabs were expected to approach billion-dollar price tags for physical plant and equipment alone.

Third, industry expansion opened up additional FPD supply sources and reduced component costs for branded notebooks, OEM assembly operations, and other product needs. Continued TFT LCD price declines helped merchant integrators and pure system integrators remain cost-competitive in final goods markets, while also freeing resources for continued technological advance. TFT LCD producers needed to continue their rapid pace of product and process technology improvements to meet the looming challenge of large-scale conversion from CRTs to FPDs in the market for home television sets. PDPs were ahead of TFT LCDs in providing many of the CRT-like product attributes that consumers valued in their home TVs, including viewing angle, brightness, and smoothness of video processing. But PDP prices remained too high to penetrate markets other than professional applications and extremely wealthy consumers.[9] The consensus among analysts and industry participants held that a "magic price" of 10,000 yen per diagonal inch (about $100) would kindle the mass market for flat TVs.[10] Subsequently, costs could be expected to decline rapidly. If PDP costs declined more rapidly than TFT costs and technology could improve, the LCD technology could be supplanted just at the threshold of the mass TV market that was supposed to have made it all worthwhile. The long journey would have ended, like the biblical odyssey of Moses, not with entering the promised land but with viewing it from a nearby mountaintop.

Many observers expect TFT LCD and PDP technologies to co-exist in the different applications best suited for each. In these scenarios, TFT LCDs would retain the market for computer monitors, for which PDP pixel sizes are too large. TFT LCDs and PDPs would compete for the mid-size TV market, and PDPs would dominate the market for the largest flat TVs. But in 2001, TFT LCDs could not be ruled out as competitors in all applications.

## THE FUTURE

The coming shift from CRTs to FPDs for TV sets assures that differentiation and cost leadership will continue to co-exist as dual strategic imperatives in the industry for many years into the future. Companies' relative capabilities to globally leverage both technology and market knowledge to innovate more rapidly than competitors will remain a critical determinant of profitability. Companies will continue to develop technological means to produce ever-larger TFT LCDs for ever-lower prices, while expanding in new markets such as desktop monitors and increasingly large television screens. Alliances and long-term contracts will play increasingly important roles in linking technological capabilities with market access, particularly considering the attractiveness of strategies to pool resources to meet the high costs of future fab generations.

The geographic concentration that characterized the earliest phases of the high-volume industry's emergence will continue to erode. The importance of leveraging knowledge creation across diverse locations will not diminish, however, but grow. Even government programs increasingly acknowledge this. The government of Singapore has declared its intention to promote the city state as the next great FPD manufacturing site, setting aside a 35-acre reserve that officials hope will accommodate up to seven fabs. The goal, according to officials, is to "increase the knowledge intensity of the industrial sector."[11] The government hopes to attract current producers to invest either on their own or in consortia to erect fifth- or even sixth-generation fabs. In the face of likely vigorous competition from a Singapore production platform, the Taiwanese and Korean governments announced an end to tariffs on imported manufacturing equipment, further acknowledging the continuing, critical role of free, open global knowledge exchange to sustaining successful industry positions.[12]

## THE PRESENT

U.S. and European companies have continued to succeed in the industry with strategies that rely heavily on alliances and that decentralize global authority and accountability to operations outside of their historic home countries. The center of gravity remains in Asia for the most successful U.S. and European companies. Senior business managers have continually transformed their organizations and roles, however, as production sites proliferated outside of Japan to other Asian centers. The manner of this transformation has appeared consistent with lessons the companies learned earlier about organizing for knowledge-driven competition. Just as these companies leveraged U.S. capabilities by vesting global leadership in Japanese organizations, they have continued to expand by cultivating a high level of local initiative and globally coordinating the internationally dispersed knowledge and capabilities that result. Headquarters' and senior managers' roles have evolved as the architects of the infrastructure needed to meet this need.

Corning's Tokyo-based Advanced Display Products Division, renamed Corning Display Technologies, coordinates among substrate R&D and production sites in the U.S., Japan, and Korea, as well between the division and Corning's core R&D organization in the U.S. Affiliates retain a high degree of autonomy, as each incorporates capabilities for marketing, production, and R&D (particularly as it relates to the co-evolution of Corning's products and those of its customers). In Korea, these capabilities are housed in the Samsung-Corning Precision Glass Co. joint venture, formed in the mid-1990s. The two companies, already partners in a CRT glass venture, had undertaken technology transfers and a shared commitment to create the new business some years earlier, in anticipation of Korea's growth as a production platform. In mid-2000, Corning announced plans to invest $320 million in Taiwan-based production facilities, with the first glass furnace to begin operations in 2003.[13] This market was initially served with imports arranged through Corning Japan K.K. In 2000, Corning Japan K.K. remained the largest producer of FPD substrates in the world.[14]

In the wake of AKT's transformation into a wholly owned subsidiary of Applied Materials, the company's Japan-based operations retained legal headquarters status, while sharing authority and responsibility with operations in Santa Clara, California, according to an allocation of functional responsibilities.

AKT's president and the leader of its management team, Kam Law, maintains his principal office in Santa Clara. Marketing and human resources functions have been added to the R&D, engineering, and manufacturing functions that have always been based there. Tetsuo Iwasaki was elected AKT's chairman after the reorganization and maintains his principal office in Japan. AKT's organization structure positions Japan as headquarters for the company's installation-based business and services. The rapid evolution of the industry has continued to demand proximity among producers, equipment makers, and materials suppliers. Consequently, AKT maintains fully capable local operations in Taiwan, Korea, and Japan, incorporating marketing, sales, and service infrastructure, leveraging distinctive capabilities and expertise with a high degree of flexibility for personnel exchanges across locations as demand warrants.[15]

In early July 2001, IBM and Toshiba revealed that the companies would dissolve their LCD manufacturing partnership, dividing its people and facilities more or less equally between them. DTI was planned to turn into a 100-percent IBM subsidiary by September. IBM announced plans to then form an alliance between IBM Japan and Chi Mei Optoelectronics of Taiwan and transfer DTI's Yasu facilities to the new venture. The alliance would continue to produce notebook and desktop computer screens, including larger-sized, extremely high-resolution, 9.2 million pixel displays based on new advances reported from IBM's Yorktown Heights lab in August 2000. IBM would license its continuing production advances to the new venture, which would receive support from Yorktown Heights and IBM Japan. DTI's Himeji facilities were to be transferred to Toshiba, which would use them as the foundation for a new venture that would use advanced new low-temperature polycrystalline technologies to produce small- and medium-sized TFT LCDs (described in more detail later in this chapter). In the years immediately preceding the change, DTI had continued to add new capacity, but in the process had increased the proportion of output dedicated to IBM. Although the DTI venture remained profitable in 2001, according to the partners, it was clear that their technology preferences for continuing and future investments had evolved in different directions.[16]

LG.Philips LCD remains headquartered in Seoul. At the end of 2000, LG and the Dutch consumer electronics giant deepened their mutual engagement by also combining their CRT operations in a new, Hong Kong-based alliance.[17]

## KNOWLEDGE-DRIVEN COMPETITION IN OTHER INDUSTRIES

Similar patterns have emerged in recent years as elements of reality in other high-technology industries. In these industries, companies commonly establish leadership of important businesses and functions outside of their home countries, while globally coordinating a combination of internationally dispersed competencies, strategic alliances, and localized initiatives for knowledge creation. Semiconductor production, manufacturing equipment and materials, wireless telecommunications, and venture capital companies that support new technology development represent salient examples.

U.S. chip producers appear to have learned lessons that eluded many U.S. FPD producers and have started to implement policies more open to global sourcing of equipment and materials as well as close production and R&D partnering with Asian and European counterparts. According to the *New York Times,* "fears of Japanese domination have abated" while, due to the growing cost of R&D and next-generation plant and equipment, "continuing advances are now beyond the grasp of Americans alone."[18] In 1999, SEMATECH, the industry consortium that originated with government support as a public/private partnership to rebuild U.S. competitiveness in chip manufacturing, consolidated with its subsidiary consortium, International SEMATECH, which had been established in 1997.[19]

Wireless telecommunications, like high-volume FPD manufacturing, represents a young industry that emerged first as a mass phenomenon outside the United States. European, Japanese, and Korean wireless telecom companies and infrastructure have significantly outpaced U.S. developments.[20] U.S. and Japanese companies have taken steps to participate directly in Europe both through alliances with the wireless giants, Nokia and Ericsson, and by establishing self-standing operations. Hewlett-Packard, IBM, Matsushita, Intel, Microsoft, Oracle, Phone.com, and Sun Microsystems are among the companies establishing wireless competence, marketing, and/or R&D centers in Finland, Sweden, or Denmark. These centers will leverage U.S. Internet leadership with Scandinavia's wireless telecom leadership in an effort to shape the new, global wireless Internet market, which has just started to emerge in both world regions. Working with Europe's leading wireless equipment manufacturers, service providers, and each other, these companies are gunning the development and market testing of new wireless Internet products such as wireless portals and microbrowsers. By injecting themselves into Europe's denser population of more advanced users, the U.S. companies gain knowledge of the likely shape of the U.S. market of the future, as well as a head start in creating the products to serve it.[21] Staying at home would risk the lead U.S. companies established in the Internet, as access migrates from the desktop to wireless handsets. Market forecasts suggest that more than two-thirds of Europeans will carry Net-enabled handsets by 2003, while lagging U.S. wireless infrastructure will keep many Americans tethered to their desks.[22]

The emerging wireless Internet has created a new industry atmosphere in Europe resembling the U.S. Internet surge of the 1990s, complete with a burgeoning wave of startups hungry for venture capital. The European venture capital industry has responded in a correspondingly free-wheeling fashion, its traditional, conservative norms transformed, in part, by knowledge accumulated through U.S. operations.[23] In the early 1990s, European technology companies could not rely on home-based venture capitalists to confidently evaluate their proposals or to provide an appropriate scale of funding for ambitious, high-risk ventures. Some, like the French FPD startup PixTech, even established themselves in the U.S., in part, to gain easier access to venture capital. PixTech's chair, Jean-Luc Grand-Clément, commented, "If U.S. venture capitalists put money into us, the Europeans will do the same."[24] Japanese startups have historically experienced even greater difficulties raising capital domestically. But in the late 1990s, it was the European and Japanese venture capitalists, rather than those seeking funding, who were setting up shop in the United States. Nokia and Siemens, for example, have based their global venture capital businesses in Silicon Valley.[25] Governments, too, have gotten into the act by setting up consular missions in Silicon Valley, San Francisco, and Boston to promote education in entrepreneurship, while networking home venture capitalists and new ventures with U.S. counterparts.[26]

## INDUSTRY CREATION AND RE-CREATION

The idea of knowledge-driven competition suggests an industry comprising companies that continually reinvent themselves. As the pace of economic change continues to accelerate, all companies will need to constantly reinvent themselves in order to survive. In this sense, all industries have transformed into knowledge-driven industries, and all companies will face a need to continually manage new industry creation.

Most FPD industry participants once subscribed to the view that the industry would eventually mature. Product innovation for differentiation would give way to a distinctive competitive phase dominated by process innovation aimed at reducing costs. Technological standards would take hold, technology-based differentiation would diminish as a mainstay of competition, and price competition would permanently dominate. Almost since the first rudimentary LCDs were commercialized in watches and calculators, participants debated whether or not producers faced a totally price-driven commodity market. Many observers believed that international diffusion of manufacturing beyond the Japanese production base would definitively lead to such a development. But only prospective U.S. producers were ever deterred by this notion.

Standards have not taken hold, and the leading edge of the display market remains a hotbed of technological advance and product differentiation. Observers regard ongoing discussions of industry standards for fifth-generation substrate sizes with skepticism. FPD producers have continued to demand new generation production equipment scaled in a variety of sizes to suit distinctive strategies for the future market.[27] As FPD technologies' leading edge moved across desktop monitors and into TV sets as profit mainstays, new entrants from new countries have taken over old product categories, such as notebook screens. These companies are recreating for themselves the knowledge basis for a leap ahead into the most advanced products and processes. New competitors from new countries and new national production platforms present the most salient challenge for the pioneering manufacturers to continue to reinvent themselves and to recreate the industry.

Many are doing so with forays into new TFT and non-TFT FPD technologies that promise to outperform current TFT LCDs, or if not, to force the technology to continue its rapid rate of improvement. Several PDP producers have declared that costs for these panels will decline sufficiently in 2002 to ignite the mass market for PDP TVs. The commercialization of field-emission display (FED) technology has been the objective of the managerially innovative PixTech global strategic alliance network, as well as of a well-funded Silicon Valley start-up, Candescent Technologies. FEDs, sometimes referred to as "flat CRTs," have long promised to provide the smooth video processing of the older technology in a scalable FPD format. FEDs' proponents have progressed far more slowly than they hoped, but the passage of time has continued to bring improvements. Most current TFT LCD producers have hedged their bets with FED research programs.

Organic electroluminescent (EL) and organic light emitting diode (OLED) displays offer another potential challenge to TFT LCDs that builds on the existing technology. These new technologies offer wider viewing angles, lower power consumption, do not require backlights, and may prove cheaper than alternative technologies to fabricate. In late 2000, the first displays using these technologies appeared in wireless telephone handsets.[28] One senior manager referred to the new OLED screens as "remarkably legible" and the technology as "a real wild card."[29]

Wireless Internet appliances have also provided an impetus to the development of polycrystalline silicon TFT LCDs, once thought too expensive to manufacture for mass market applications because of the high cost of substrate materials needed to withstand the necessary high processing temperatures. Technology improvements have increased the temperatures that glass substrates can withstand in fabrication processes, while process improvements have lowered the temperatures necessary for fabricating polycrystalline TFTs. Low-temperature polycrystalline, or LTP TFTs, offer potential advantages in resolution and permit driver chips to be simultaneously etched on glass substrates in the TFT manufacturing process, saving physical space. Small TFT LCDs manufactured in this way perform ideally as color wireless handset displays, for which wireless Internet developments will ignite immense demand. Like the calculator and watch displays of the early 1970s, these small displays may create the knowledge foundations for a technology evolution that will directly challenge the incumbent technology, be it amorphous silicon TFT LCDs, PDPs, or a renascent CRT. Toshiba has taken the lead in LTP, reporting 70-percent yield on its production line in September 2000. Taiwan's Prime View International announced its intention to begin LTP TFT LCD manufacturing in 2001.[30]

More than thirty years have passed since the first inventions that directly led to the current, dominant amorphous silicon TFT LCD technology for FPDs. Ten years have passed since the high-volume industry took off. The FPD industry shows no signs of moving beyond its creation phase. The details of management challenges remain the same, but the intensity and stakes have grown higher. The industry continues to attract new entrants, and incumbent companies continue to add capacity, resulting in unremitting price erosion. In many consumers' minds, FPDs have become inseparably intertwined with high-definition television (HDTV). TFT LCD technology has continued improving on attributes that consumers associate with HDTV, particularly high resolution, speedy video response rates, and wide viewing angle. Other flat panel display technologies—especially PDPs—are doing the same for larger screen TVs.

Sometime within the next ten years, decreasing costs for large TFT LCDs and PDPs (or some other technology waiting in the wings), widespread availability of HDTV programming, and CRT-like video quality will converge to create yet another FPD industry boom, by far the most astonishing ever. The absolute costs of entry will have ballooned two- to four-fold, beyond the resources of most individual companies. Alliances will pervade to an even greater extent than at the turn of the millennium.

The FPD industry of 2010 will make the industry of 2000 seem very much like the sleepy little Ohio River Valley town of the 1890s that Shinichi Hirano imagined it once to be.

There will be new wild ducks and young engineers. But there will also be nostalgia for the shared sense of challenge that went before actually achieving things that many regarded as impossible, or at best, unwise.

As for now, there remain a few sparsely settled lands to the west.

# APPENDIX ONE

**1888**

Friedrich Reinitzer, an Austrian botanist, discovers the unique physical properties of liquid crystal.

**1897**

Karl Ferdinand Braun invents the cathode ray tube.

**1937**

IBM Japan established.

**1947**

December 16: Walter Brattain, John Bardeen, Robert Gibney, and William Shockley invent the transistor at AT&T's Bell Labs.

**1960**

Richard Williams demonstrates at RCA's Sarnoff Research Center that a liquid crystal substance in its transparent state turns opaque, scattering (or reflecting) light instead of transmitting it, when charged with an electric current.

**1964**

George Heilmeier begins to lead experiments at Sarnoff that eventually harness Williams' 1960 discovery to create an image-capable display, known as the dynamic scattering LCD.

Donald Bitzer and H. Gene Slottow co-develop the alternating current plasma display panel (AC PDP) with Bitzer's graduate student, Robert Willson, at the University of Illinois' Computer-based Education Research Lab.

**1967**

T. Peter Brody and Derrick Page experiment with using tellurium films to create thin-film transistors in their lab at Westinghouse.

**1968**

May 28: RCA introduces the dynamic scattering liquid crystal display (LCD) at a press conference in New York City.

T. Peter Brody and Derrick Page develop an active matrix electroluminescent display that uses ferro-electric switches at Westinghouse.

**1969**

James Fergason invents the twisted nematic liquid crystal display (TN LCD) while on the faculty at Kent Sate University. (Wolfgang Helfrich and Martin Schadt of Hoffman LaRoche in Switzerland arrived at the same idea independently and published a paper in 1971. See Bob Johnstone, *We Were Burning*, 108.)

**1970**

Sharp licenses dynamic scattering LCD technology from RCA instead of building a CRT factory.

**1971**

T. Peter Brody develops an active matrix liquid crystal display for the U.S. Air Force.

IBM leverages results from a joint program with Owens Illinois based on the work of Donald Bitzer's lab at the University of Illinois to demonstrate a postage-stamp-sized PDP.

**1973**

April: Sharp introduces the first LCD hand calculator, the EL-805.

October: Seiko Epson introduces the first digital LCD watch, the 06LC.

IBM begins to manufacture small monochrome PDPs in Fishkill, New York.

T. Peter Brody's Westinghouse team demonstrates applicability of TFTs for active matrix LCD drive systems.

**1974**

Paul Alt and Peter Pleshko publish the "Iron Law of Multiplexing," characterizing some of the limitations of passive matrix liquid crystal displays.

RCA ends its LCD program.

**1975**

IBM switches emphasis to larger PDPs and constructs a new plant in Kingston, New York.

**1976**

Westinghouse closes its electron tube division.

**1978**

Westinghouse closes down its research project on active matrix LCDs headed by T. Peter Brody.

**1979**

Walter Spear and Peter LeComber, faculty members of the University of Dundee in Scotland, publish a paper that demonstrates that TFTs could be fabricated out of amorphous silicon. The paper also demonstrates the feasibility of using amorphous silicon with liquid crystal to make displays.

**Early 1980s**

Satoshi Furuyama, working in sales for Corning Japan, follows a paper trail of unexplained orders for advanced glass products that puts the company in touch with early R&D efforts in Japan aimed at commercializing TFT LCDs.

**1982**

April 14: Grid Systems introduces a portable computer that weighs 9.5 pounds, priced at $8,150.

Shinji Morozumi and Koichi Oguchi light up a tiny TFT LCD television in their lab at Seiko Suwa.

**1983**

May: Seiko Suwa demonstrates a tiny prototype of the first color TFT LCD TV at the annual meeting of the Society for Information Display.

IBM introduces 17-inch monochrome PDPs. Most are sold to the financial services industry.

IBM convenes the first of three task forces to consider alternative FPD technologies and their likely impact on IBM's businesses. The task force recommends shutting PDP research and production down.

IBM shuts down plasma research.

(approximate) Corning retools a fusion glass facility threatened with closure in Harrodsburg, Kentucky, for TFT LCD glass substrate production.

(approximate) Corning begins producing its first generation 7059 fusion-formed borosilicate glass substrates for TFT LCDs at Harrodsburg.

Tektronix spins off Planar Systems, Inc., a U.S. manufacturer of EL displays.

**1984**

Seiko Epson markets its 2-inch TFT LCD TV, manufactured using polycrystalline silicon on quartz substrates, for $315.

**1986**

May: Matsushita introduces a 3-inch TFT LCD portable color television manufactured using amorphous silicon.

August 1: IBM and Toshiba contract to conduct joint R&D toward a very large TFT LCD prototype. The project team begins work.

Sharp begins high-volume production of STN LCDs for notebook computers.

IBM shuts down PDP manufacturing.

**1987**

August: Plasmaco is co-founded by Larry Weber and purchases the manufacturing equipment from IBM's Kingston PDP facility.

Sharp introduces an early personal digital assistant, known as the Wizard, using an STN LCD.

Sharp begins high-volume manufacturing of 3-inch TFT LCDs for small TVs.

**1988**

June 17: Sharp announces its development of a 14-inch color TFT LCD.

September 21: IBM and Toshiba announce the joint development of a 14.26-inch color TFT LCD.

Sharp begins high-volume manufacturing of 4-inch TFT LCDs.

Corning opens additional TFT LCD glass melting capacity as well as finishing facilities in Shizuoka, Japan.

**1989**

August 30: IBM and Toshiba announce plans to establish DTI, a TFT LCD manufacturing alliance.

November 1: DTI officially starts up.

The Corning substrates business moves from the Technical Products Division to the new Advanced Display Products Division of the Information Display Group.

Applied Materials responds to Toshiba's first request for a CVD proposal, but the customer finds the proposed equipment more technically sophisticated than necessary.

## 1990

OIS, Planar, Plasmaco, Photonics Technologies, Magnascreen, Cherry
Corporation, and Electroplasma form the Advanced Display Manufacturers
of America (ADMA)

July 17: ADMA's members file an anti-dumping petition with the U.S.
Department of Commerce against thirteen Japanese display manufacturers.

August: NEC starts up the first Generation 1 TFT LCD fabrication line.

## 1991

May 15: DTI begins producing large-format color TFT LCDs using
Generation 1 equipment.

August 15: The U.S. International Trade Commission authorizes anti-dumping
duties on several Japanese FPD companies' products.

Fall: Sharp beings manufacturing large-format color TFT LCDs using
Generation 1 equipment.

With its earlier CVD proposal to Toshiba revived and other potential customers interested, Applied Materials creates a new business unit, Applied
Display Technology (ADT), to build TFT LCD manufacturing equipment.
ADT locates its headquarters in the Kansai region of Japan near its key
customers' fabs.

## 1992

November: IBM introduces the ThinkPad Model 700C, the first hit notebook
computer with a color TFT LCD display. Industry-wide shortages of TFT
LCDs ensue almost immediately.

November: Optical Imaging Systems (OIS) requests that the U.S. Commerce
Department remove the anti-dumping duties on TFT LCDs.

## 1993

ADT ships its first CVD prototypes to beta sites on Gen 1 lines at DTI and
Sharp.

Microsoft introduces Windows 3.1 operating system.

June 17: Applied Materials and Komatsu enter into a strategic alliance to
manufacture and market TFT LCD manufacturing equipment. The alliance
is called Applied Komatsu, Inc. (AKT).

June 21: The U.S. Department of Commerce drops anti-dumping duties imposed on TFT LCD imports from Japan.

July 20: The United States Display Consortium is founded with a $20-million grant from DARPA to be matched by U.S. industry members on a project-by-project basis.

October: AKT announces its first commercial CVD tool, the AKT-1600, available on a six-month order turnaround.

## 1994

May: Sharp starts up the first second-generation TFT production line.

June: DTI starts up a second-generation line; Sharp starts up an additional second-generation line.

April 26: The Clinton administration announces the National Flat Panel Display Initiative.

June: Plasmaco exhibits a working AC PDP at the SID Annual Meeting.

December: Hitachi starts up second-generation TFT LCD production.

## 1995

January 17: A devastating earthquake strikes the Kansai region of Japan.

February: Samsung begins TFT LCD production on its Gen 2 fab in Kiheung.

March: Fire damages the new OIS TFT LCD fab in Northville, Michigan.

July: LG begins TFT LCD production on its Gen 2 fab in Kumi.

July: Samsung achieves commercial yields on its first high-volume line, a Gen 2

July: Sharp begins production at new Gen 2.5 and Gen 3 fabs in Mie, Japan.

Fall: DTI begins production at its Gen 3 fab in Yasu, Japan.

Samsung-Corning Precision Glass opens fusion glass melting and finishing facilities in Kumi.

## 1996

January 9: Matsushita acquires Plasmaco. Larry Weber remains president and CEO.

February 12: Xerox announces that it will manufacture high-resolution TFT LCDs for its own use and for the high-end workstation and graphic systems markets.

March: Lucent Technologies announces that it will shut down its TFT LCD R&D.

October: The USDC and Korea's EDIRAK sign a memorandum of under-standing to jointly support development of FPD manufacturing equipment by funding projects that partner U.S. equipment and materials suppliers with Korean FPD producers.

October: Samsung starts up Generation 3 fab.

October 14: Philips and Hosiden announce that they have agreed to form a strategic alliance.

**1997**

Financial crisis begins to grip Asia.

Early in the year, Samsung achieves commercial yields on its Gen 3 line.

March 17: The Hosiden and Philips Flat Panel Display Company begins operations.

**1998**

Applied Materials and Komatsu end AKT alliance. Applied Materials will carry the business forward on its own.

September 18: OIS announces plans to shut down.

Sharp announces the first complete line of flat TVs.

**1999**

March 12: The U.S. TFT LCD maker dpiX, a Xerox Company, informs staffers of a sixty-day wind-down period prior to closing, contingent on the outcome of efforts to find a buyer.

May 17: A European consortium including Philips, Siemens, and Thomson Electronics together with Planar Systems and Varian Medical System announce that they will acquire 80 percent of dpiX. Xerox will continue to hold a 20-percent ownership stake in the company, the last large-format TFT LCD manufacturing facility remaining open in the U.S.

September: LG.Philips joint venture begins operations, following a May announcement that Philips would invest $1.6 billion in exchange for a 50-percent interest in LG's TFT LCD business.

The USDC alters its charter to admit non-U.S. owned companies to associate membership.

Samsung and LG.Philips capture the first- and second-market share ranks in large-format color TFT LCDs.

**2000**

Plasma TV sets begin to penetrate high-end consumer market in significant numbers. Senior managers in several companies declare that PDP will break through the price barrier to mass market penetration by 2002.

# APPENDIX TWO

**Table 1** 1990–2000 Worldwide Flat Panel Display Sales;
Active and Passive Matrix Liquid Crystal (AMLCD and PMLCD) Display
Total Sales in Billions of U.S. Dollars and Percent of Worldwide Market

| Year | Total Sales | AMLCD Sales[1] | AMLCD % | PMLCD Sales | PMLCD %[2] |
|------|-------------|----------------|---------|-------------|------------|
| 1990 | 3.17 | 0.22 | 7.0 | 2.89 | 91.0 |
| 1991 | 3.87 | 0.53 | 14.0 | 3.28 | 84.0 |
| 1992 | 4.11 | 0.92 | 22.0 | 3.13 | 77.0 |
| 1993 | 5.79 | 2.16 | 37.0 | 3.55 | 62.0 |
| 1994 | 8.56 | 4.11 | 48.0 | 4.33 | 50.6 |
| 1995 | 8.02 | 4.05 | 50.6 | 3.84 | 48.0 |
| 1996 | 9.66 | 5.91 | 61.0 | 3.59 | 37.0 |
| 1997 | 10.46 | 6.58 | 63.0 | 3.60 | 34.0 |
| 1998 | 9.98 | 6.67 | 67.0 | 2.91 | 29.0 |
| 1999 | 21.43 | 13.35 | 62.0 | 3.57 | 16.0 |
| 2000 | 29.47 | 16.73 | 57.0 | 4.64 | 16.0 |

SOURCE: DisplaySearch.
[1]Mainly TFT LCDs.
[2]Mainly STN LCDs.

**Table 2** 1990–2000 Worldwide Flat Panel Display Sales; Plasma Display Panel
(PDP) and Electroluminescent Display (EL) Sales in Billions of U.S. Dollars and
Percent of Worldwide Market

| Year | Total Sales | PDP Sales | PDP % | EL Sales | EL % |
|------|-------------|-----------|-------|----------|------|
| 1990 | 3.17 | – | – | 0.06 | 2.0 |
| 1991 | 3.87 | – | – | 0.06 | 2.0 |
| 1992 | 4.11 | – | – | 0.06 | 1.0 |
| 1993 | 5.79 | – | – | 0.08 | 1.0 |
| 1994 | 8.56 | 0.03 | 0.4 | 0.09 | 1.0 |
| 1995 | 8.02 | 0.03 | 0.4 | 0.10 | 1.0 |
| 1996 | 9.66 | 0.05 | 0.5 | 0.11 | 1.0 |
| 1997 | 10.46 | 0.16 | 2.0 | 0.12 | 1.0 |
| 1998 | 9.98 | 0.27 | 3.0 | 0.14 | 1.0 |
| 1999 | 21.43 | 3.10 | 14.5 | 1.41 | 6.5 |
| 2000 | 29.47 | 6.53 | 22.0 | 1.57 | 5.0 |

SOURCE: DisplaySearch.

**Table 3** 1996–2000 Rank Order of Large Format (greater than 10.4-inch)
TFT LCD Manufacturers

| 1996 | 1997 | 1998 | 1999 | 2000 |
|------|------|------|------|------|
| DTI | DTI | DTI/Samsung[1] | Samsung | Samsung |
| Sharp | Sharp | Sharp | LG.Philips | LG.Philips |
| NEC | NEC | NEC | DTI | Hitachi |
| Hitachi | Samsung | LG | Sharp | DTI |
| | Hitachi | Hitachi | Hitachi | Sharp |
| | | | NEC | NEC |

SOURCE: Authors' calculations based on various industry sources including interviews.

[1]Industry sources are equivocal regarding the relative rankings of DTI and Samsung in 1998.

**Table 4** 1990–1998 Market Leaders by AMLCD
Revenues, All Sizes; 1990–2000 Market Leaders,
Large-Format TFT LCD Unit Sales[1]

| Year | Revenues, All Sizes | Large-Format Units |
|------|---------------------|--------------------|
| 1990 | Sharp | |
| 1991 | Sharp | NEC |
| 1992 | Sharp | Sharp |
| 1993 | Sharp | Sharp |
| 1994 | Sharp | Sharp |
| 1995 | Sharp | Sharp |
| 1996 | Sharp | DTI |
| 1997 | Sharp | DTI |
| 1998 | Sharp | DTI/Samsung[2] |
| 1999 | | Samsung |
| 2000 | | Samsung |

SOURCE FOR REVENUES: DisplaySearch; source for units: authors'
calculations based on various industry sources including interviews.

[1]Prior to 1996, greater than 8.4-inch TFT LCD; from 1996, greater than
10.4-inch TFT LCD.

[2]Industry sources are equivocal regarding the relative rankings of DTI and
Samsung in 1998.

**Table 5** 1991–2000 Color TFT LCD Prices in U.S. Dollars by Size (diagonal inches)

| | | 8.4" | 10.4" VGA | 12.1" SVGA | 13.3" XGA | 14.1" XGA |
|---|---|---|---|---|---|---|
| 1991 | Q1 | 3,700 | | | | |
| | Q3 | | 3,000 | | | |
| 1992 | Q1 | | 2,500 | | | |
| | Q3 | | 2,000 | | | |
| 1993 | Q1 | | | | | |
| | Q2 | 800–900 | | | | |
| | Q3 | | 1,500 | | | |
| | Q4 | | | | | |
| 1994 | Q1 | | | | | |
| | Q2 | 800–1,300 | 1,200 | | | |
| | Q3 | | | | | |
| | Q4 | | 1,000 | | | |
| 1995 | Q1 | | | | | |
| | Q2 | | 952 | | | |
| | Q3 | | 811 | | | |
| | Q4 | | 476 | | | |
| 1996 | Q1 | | 300 | 1,450–852 | | |
| | Q2 | | 370 | 830 | | |
| | Q3 | | 376 | 735 | | |
| 1997 | Q4 | | | 740 | | |
| | Q1 | | | 650 | 1,050 | 1,400 |
| | Q2 | | | 580 | 950 | 1,150 |
| | Q3 | | | 480 | 850 | 950 |
| | Q4 | | | 430 | 650 | 750 |
| 1998 | Q1 | | 285 | 350 | 550 | 650 |
| | Q2 | | 204 | 260 | 215 | 500 |
| | Q3 | | 180 | 225 | 360 | 410 |
| | Q4 | | 180 | 225 | 350 | 395 |
| 1999[1] | Q1 | | | 300 | 430 | 460 |
| | Q2 | | | 340 | 485 | 520 |
| | Q3 | | | 360 | 505 | 535 |
| | Q4 | | | 378 | 519 | 547 |
| 2000[2] | Q1 | | 300 | 385 | 515 | 545 |
| | Q2 | | | 355 | 425 | 495 |
| | Q3 | | | 310 | 390 | 440 |
| | Q4 | | 230 | 260 | 340 | 395 |

SOURCES: DisplaySearch, 1997–2000, except for 2000 10.4-inch; authors' research including various media and interview materials, 1991–1996 and 2000 10.4-inch. XGA or fully extended graphics array compatible displays consist of 1024 by 768 pixels, while super-video graphics array or SVGA have 800 by 600 pixels. Display resolution increases with the number of pixels.

[1]End quarter prices.

[2]Highest prices in a range given at end quarter.

# APPENDIX THREE

This appendix presents an index of people, their titles, and affiliations at the time the authors formally interviewed them for this book. We have listed individuals in alphabetical order according to surname, but written each entry starting with the interviewee's given name. In many cases, the formal meetings documented here served as introductory encounters in relationships that have grown over time through repeated contact. We deeply appreciate the time and thought each of the people named here invested in our project. We informally encountered many other executives, scientists, engineers, and officials in the course of our study, as well as a few who wished to remain anonymous, to whom we are also grateful.

**ITSURO ADACHI,** general manager, Color LCD Division, Display Devices Operations Unit, NEC Corporation, Tokyo, Japan, June 10, 1997.

**MASAO AMANO,** senior manager of information systems, Display Technologies, Inc. (DTI), Himeji, Japan, June 2, 1997.

**TOSHIMASA ASAKA,** director, member of the board in charge of Corporate Planning Office, Matsushita Electronics Corporation, Osaka, Japan, November 1, 1996.

**DAVID L. BERGERON,** vice president, manufacturing technology, Candescent Technologies, Inc., San Jose, California, June 27, 1996.

**JOHAN BERGQUIST,** flat panel technology analyst, Asian Technology Information Program, Tokyo, Japan, October 22, 1996.

**EDWARD J. BOLING,** vice president, research and development, Image-Quest Technologies, Fremont, California, June 25, 1996.

**WILLIAM F. BRINKMAN,** physical sciences research vice president, Bell Labs, Lucent Technologies, Murray Hill, New Jersey, July 19, 1996.

**THOMAS S. BUZAK,** president, Technical Visions Inc., Beaverton, Oregon, June 11, 1996.

**KUANG-LANG (WOLF) CHEN,** manager, Device Design Division, Department of Opto-Electronics, Chunghwa Picture Tubes, Ltd., Padeh City, Taiwan, May 30, 1997.

**B.D. CHOI,** vice president, LCD Division, Hyundai Electronics Industries Co., Ltd., Ichon, Korea, December 9, 1996.

**MICHAEL F. CIESINSKI,** chief executive officer, U.S. Display (USDC), San Jose, California, July 2, 1996.

**FRANCIS G. COURREGES,** executive vice president, PixTech, Montpellier, France, December 20, 1996.

**THOMAS L. CREDELLE,** director, product marketing, Automotive, Energy and Controls Group, Flat Panel Display Division, Tempe, Arizona, March 27, 1997.

**STEVEN W. DEPP,** director, Subsystem Technologies and Applications Laboratory, International Business Machines Corporation (IBM), Thomas J. Watson Research Center, Yorktown Heights, New York, July 22, 1996; Ann Arbor, Michigan, January 24, 1997, March 21, 1997.

**ROBERT M. DUBOC, JR.** executive vice president and COO, Candescent Technologies, Inc., San Jose, California, July 1, 1996.

**FARID DURRANI,** director, LCD manufacturing operations, LCD Division, Sharp Microelectronics Technology, Inc., Camas, Washington, June 12, 1996.

**THEODORE S. FAHLEN,** vice president, research and development, Candescent Technologies, Inc., San Jose, California, June 27, 1996.

**LISA FERRERO,** marketing analyst, Corning Japan, K.K., Tokyo, Japan, November 7, 1996.

**KEN FLAMM,** senior fellow, Foreign Policy Studies Program, The Brookings Institution, Washington, D.C., December 18, 1995.

**GERALD J. FINE,** deputy general manager, Advanced Display Products, Corning, Incorporated, Corning, New York, September 13, 1996.

**SATOSHI FURUYAMA,** president, Corning Japan, K.K., Tokyo, Japan, November 7, 1996.

**JEAN-LUC GRAND-CLÉMENT,** chairman and CEO, PixTech, Mountain View, California, June 25, 1996.

**LARRY GRAVES,** director, roadmap and standards, U.S. Display Consortium (USDC), San Jose, California, June 24, 1996.

**DONALD HARRIS,** marketing manager, ImageQuest Technologies, Inc., Fremont, California, June 25, 1996.

**MARK A. HARTNEY,** program manager, display manufacturing technology, United States Department of Defense, Advanced Research Projects Agency, Washington, D.C., December 18, 1995.

**TAKAHISA HASHIMOTO,** director, Display Technology, Asia Pacific Technology Operations (APTO), IBM Japan, Ltd., Yamato, Japan, November 6, 1996.

**KATSUNORI HATANAKA,** manager, FV Project-I, Canon Inc. R&D Headquarters, Canon Research Center, Atsugi, Japan, November 5, 1996.

**AL HERMAN,** vice president and general manager, Planar Advance, Inc., Beaverton, Oregon, June 10, 1996.

**SHINICHI HIRANO,** IBM senior technical staff member, Display Technology, IBM Japan, Ltd., Yamato, Japan, November 6, 1996.

**SCOTT HOLMBERG,** president and CEO, ImageQuest Technologies, Inc., Fremont, California, June 25, 1996, and Minneapolis, Minnesota, August 1, 1996.

**TOM HOLZEL,** vice president, sales and marketing, PixTech, Mountain View, California, June 25, 1996.

**CHUNG T. HO,** display process development, Plasmaco, Inc., Highland, New York, July 24, 1996.

**DAISUKE HONGU,** manager, Planning Group, Corporate Planning Office, Matsushita Electronics Corporation, Osaka, Japan, November 1, 1996, and June 6, 1997.

**WEBSTER E. HOWARD,** director, high resolution technologies, Global Manufacturing and Engineering, AT&T, Murray Hill, New Jersey, April 11, 1996; also as vice president and chief technology officer, FED Corporation, Hopewell Junction, New York, July 23, 1996.

**DAVID C. HSING,** senior advisor, Industrial Technology Research Institute (ITRI), Hsinchu, Taiwan, May 28, 1997.

**DYI-CHUNG HU,** senior director, R&D and LCD Manufacturing Division, Prime View International, Hsinchu, Taiwan, May 27, 1997.

**JAMMY CHIN-MING HUANG,** Vacuum Microelectronics Development Department, Deep Submicron Division, Industrial Technology Research Institute, Electronics Research and Service Organization (ITRI/ERSO), Hsinchu, Taiwan, May 28, 1997.

**RONALD L. HUFF,** vice president, process development, ImageQuest Technologies, Inc., Fremont, California, June 25, 1996.

**DENNIS A. HUNTER,** managing director, corporate development, Applied Materials, Santa Clara, California, June 24, 1996.

**JAMES M. HURD,** president/CEO, Planar Systems, Inc., Beaverton, Oregon, June 10, 1996.

**DOUG H. HWANG,** assistant manager, LCD North America Sales Team, LG Electronics, Inc., Seoul, Korea, December 6, 1996.

**SHOICHI IINO,** general manager, LCD Development Center, Seiko Epson Corporation, June 13, 1997.

**HIDENORI IKEDA,** senior manager (color LCDs, FIPs), 1st Sales Promotion Department, Electronic Component Marketing Division, NEC Corporation, Tokyo, Japan, November 7, 1996, and June 10, 1997.

**KIYOKAZU IMANISHI,** manager, planning section, Planning Department, Liquid Crystal Display Division, Matsushita Electric Industrial Co., Ltd., June 6, 1997.

**YASUKI IWANO,** manager, marketing, Hosiden and Philips Display Corp., Kobe, Japan, June 3, 1997.

**TETSUO IWASAKI,** president, CEO, Applied Komatsu Technology, Inc. (AKT), Kobe, Japan, October 24, 1996; Tokyo, Japan, June 9, 1997.

**GARY W. JONES,** president/CEO, FED Corporation, Hopewell Junction, New York, July 23, 1996.

**SUSAN K. JONES,** executive vice president, FED Corporation, Hopewell Junction, New York, July 23, 1996.

**CHARLES W. (CHIP) JONES,** product line manager, Corning Japan, K.K., Tokyo, Japan, November 7, 1996.

**GORDON JORGENSON,** information processing and displays, Honeywell Technology Center, Honeywell, Inc., Minneapolis, Minnesota, December 20, 1995.

**THOMAS A. KALIL,** director to the National Economic Council, The White House, Washington, D.C., December 18, 1995.

**KEINOSUKE KAMADA,** president and representative director, Hosiden and Philips Display Corp., Kobe, Japan, June 3, 1997.

**TORU KANEKO,** manager, LCD Business Planning Department, LCD Division, Seiko Epson Corporation, Tokyo, Japan, November 8, 1996.

**TADAMICHI KAWADA,** director, corporate marketing, Applied Komatsu Technology, Inc. (AKT), June 9, 1997.

**TSUYOSHI KAWANISHI,** senior advisor, Toshiba Corporation, Tokyo, Japan, November 5, 1996.

**HIROAKI KAWASAKI,** director, product planning, Futaba Corporation, Tokyo, Japan, October 21, 1996.

**KOTARO KAZUTA,** Planning Department, TFT-LCD Development Center, Tenri Liquid Crystal Display Group, Sharp Corporation, Tenri, Japan, June 6, 1997.

**JIN-HYUN KIM,** president, Seoul City University, Seoul, Korea, December 7, 1996.

**JOO-YOUL KIM,** manager, LCD sales, Samsung Display Devices, Seoul, Korea, December 10, 1996.

**O.H. (JOHN) KIM,** manager, Marketing Department, LCD Division, Hyundai Electronics Industries Co., Ltd., Ichon, Korea, December 9, 1996.

**WOO-YEOL KIM,** research fellow, Products Engineering Department, LCD Division, LG Electronics, Inc., December 6, 1996.

**HIROSHI KIMURA,** senior executive vice president, Display Technologies, Inc. (DTI), Himeji, Japan, June 2, 1997.

**RICK KNOX,** director, advanced technology, Corporate Development, Compaq Computer Corporation, Houston, Texas, March 26, 1997.

**HIROSHI KOJIMA,** general manager, LCD Business Planning Department, LCD Operations Division, Seiko Epson Corporation, Suwa, Japan, June 13, 1996.

**HIROSHI KOMATSU,** general manager, LCD Business Planning Department, LCD Division, Seiko Epson Corporation, Tokyo, Japan, November 8, 1996.

**JA-POONG KOO,** general manger, Electronic Display Industrial Research Association of Korea (EDIRAK), Seoul, Korea, December 2, 1996.

**TAKASHI KURUMISAWA,** chief engineer, LCD Development Center, Seiko Epson Corporation, Suwa, Japan, June 13, 1997.

**MASAYUKI KURIMOTO,** standards senior coordinator, Japan Office, Semiconductor Equipment and Materials International (SEMI), Tokyo, Japan, October 21, 1996.

**DAVID R. LAMB,** associate director of research, information processing and displays, Honeywell Technology Center, Honeywell, Inc., Minneapolis, Minnesota, December 20, 1995.

**CHOON-RAE LEE,** managing director, LCD Division, LG Electronics, Inc., Seoul, Korea, December 6, 1996.

**D.S. LEE,** manager, LCD Sales/North America, LG Electronics, Inc., Seoul, Korea, December 6, 1996.

**JAMES H. LEE,** manager, USA LCD Export Team, Samsung Display Devices, Seoul, Korea, December 10, 1996.

**JANG-JENG LIANG,** director, Department of Opto-electronics, LCD plant manager, Chunghwa Picture Tubes, Ltd., Padeh City, Taiwan, May 30, 1997.

**JAE-CHOON LIM,** director general for machinery and electronics research coordination, Ministry of Science and Technology, Republic of Korea, Kwacheon, Korea, December 4, 1996.

**SUNGKYOO LIM,** consultant, Electronic Display Industrial Research Association of Korea (EDIRAK), and professor, Dankook University Information Display Research Center, Seoul, Korea, December 2, 1996.

**TOM LONG,** director of programs, Planar Advance, Inc., Beaverton, Oregon, June 10, 1996.

**IAN MACKAY,** director, Display Technology Systems, St. Paul, Minnesota, September 11, 1996.

**HARRY A. MARSHALL,** chairman, president and CEO, Candescent Technologies, Inc., San Jose, California, July 1, 1996.

**JAMES C. MCGRODDY,** senior vice president, research, (retired), International Business Machines Corporation (IBM), Armonk, New York, October 27, 1997.

**ANDY MEYERS,** special advisor, Embassy of the United States of America, Tokyo, Japan, October 18, 1996.

**AKIHISA MITSUHASHI,** leader, sales and marketing, commercial operations, Corning Japan, K.K., Tokyo, Japan, November 7, 1996.

**WALTER F. MONDALE,** Ambassador of the United States of America to Japan, Tokyo, Japan, October 18, 1996.

**SHINJI MOROZUMI,** deputy general manager, ALCD Business Division, Hosiden Corporation, Kobe, Japan, October 25, 1996.

**SAMUEL A. MUSA,** executive director, Center for Display Technology and Manufacturing, University of Michigan, Ann Arbor, Michigan, January 24, 1997.

**SOICHI NAGAMATSU,** director, Industrial Electronics Division, Machinery and Information Industries Bureau, Ministry of International Trade and Industry (MITI), Tokyo, Japan, October 22, 1996.

**KEISHI NARUMI,** manager, LD Business Planning Department, LCD Division, Seiko Epson Corporation, Suwa, Japan, June 13, 1997.

**NORIHIKO NAONO,** director of business development, Rambus, K.K., Tokyo, Japan, October 17, 1996.

**KOHKI NODA,** program manager, emerging technology, Display Technology, IBM Japan, Ltd., Yamato, Japan, November 6, 1996.

**TAKUYA OGAWA,** specialist, Information Technology Unit, Embassy of the United States of America, Tokyo, Japan, June 12, 1997.

**KYE HWAN OH,** executive vice president, Display Devices and Electronic Components, Hyundai Electronics Industries Co., Ltd., Ichon, Korea, December 9, 1996.

**OSAMU OKUMURA,** chief engineer, LCD Development Center, LCD Operations Division, Seiko Epson Corporation, Suwa, Japan, June 13, 1997.

**SEIKA OTOGURO,** manager, Planning Group, Production Administration Department, Liquid Crystal Display Division, Toshiba Corporation, Tokyo, Japan, June 12, 1997.

**JOSEPH L. PAONE,** manager, product line, display devices, Raytheon Electronic Systems, Quincy, Massachusetts, September 25, 1996.

**RAJENDRA I. (RAJ) PATEL,** laboratory manager, Liquid Crystal Display Program, 3M Life Sciences Sector Laboratories, St. Paul, Minnesota, September 10, 1996.

**M. ROBERT PINNEL,** chief technical officer, U.S. Display Consortium (USDC), San Jose, California, July 2, 1996.

**JOEL POLLACK,** senior product marketing manager for displays, Sharp Microelectronics Technology, Inc., Camas, Washington, June 12, 1996.

**ROBERT J. PRESSLEY,** founder and technical consultant, Candescent Technologies, Inc., San Jose, California, June 27, 1996.

**ANDREW J. QUINN,** economic officer, trade, Embassy of the United States of America, Tokyo, Japan, October 18, 1996.

**GRIFFITH L. RESOR III,** president, MRS Technologies, Inc., Chelmsford, Massachusetts, August 22, 1997.

**JOHN RITSKO,** senior manager for flat panel display technology, International Business Machines Corporation (IBM), Thomas J. Watson Research Center, Yorktown Heights, New York, July 22, 1996.

**KUNIHIRO SAKAI,** manager, Development Dept. 15, FV Project, Canon, Inc. R&D Headquarters, Canon Research Center, Atsugi, Japan, November 5, 1996.

**KANRO SATO,** general manager, Liquid Crystal Display Division, Toshiba Corporation, Tokyo, Japan, June, 12, 1997.

**ANDREW H. SHIH,** product marketing manager, flat panel display, Lam Research, Fremont, California, June 26, 1996.

**TORU SHIMA,** president, Display Technologies, Inc. (DTI), Himeji, Japan, June 2, 1997.

**MASARU SHIOZAKI,** senior manager, planning, Display Technologies, Inc. (DTI), Himeji, Japan, June 2, 1997.

**PETER SHINYEDA,** vice president and general manager, Automotive, Energy and Controls Group, Flat Panel Display Division, Motorola, Tempe, Arizona, March 27, 1997.

**JON A. SHROYER,** president and CEO, Sharp Microelectronics Technology, Inc., Camas, Washington, June 12, 1996.

**DAVID E. SLOBODIN,** program manager, display technology, United States Department of Defense, Advanced Research Projects Agency, Washington, D.C., December 19, 1995.

**MARKO M.G. SLUSARCZUK,** director of business development, Candescent Technologies, Inc., San Jose, California, June 27, 1996.

**JULIE SNYDER,** commercial attaché, Embassy of the United States of America, Tokyo, Japan, October 18, 1996, and June 12, 1997.

**HOE-SUP SOH,** general manager, LCD Division R&D Laboratory, LG Electronics, Inc., Seoul, Korea, December 6, 1996.

**VINCENT SOLLITTO,** chief executive officer and president, Photon Dynamics, Inc., San Jose, California, October 17, 1997.

**TERRY J. SONNONSTINE,** business development director, Liquid Crystal Display Program, 3M Life Sciences Sector Laboratories, St. Paul, Minnesota, September 10, 1996.

**JUN H. SOUK,** executive director, AMLCD R&D, Semiconductor Business, Samsung Electronics Co., Kiheung, Korea, December 5, 1996.

**CURTIS M. STEVENS,** executive vice president and general manager, Planar America, Inc., June 10, 1996.

**NICHOLAS G. STURIALE,** director of marketing, Candescent Technologies, Inc., San Jose, California, June 27, 1996.

**J.C. STUVE, PRESIDENT,** CEO, Philips Flat Panel Display Co., Eindhoven, The Netherlands, September 11, 1997.

**ZENZO TAJIMA,** project manager, Super TFT Project, Electron Tube and Devices Division, Hitachi, Ltd., Makahari, Japan, November 8, 1996, and Tokyo, Japan, June 9, 1997.

**TATSUO TAKAHASHI,** corporate specialist, technical, EIC Group Technical, Sumitomo 3M Limited, Tokyo, Japan, June 10, 1997.

**SEISAKU TAKATA,** manager, technical marketing, Sales and Marketing, Applied Komatsu Technology, Inc. (AKT), October 24, 1996.

**HIROSHI TAKE,** general manager, TFT 1st Engineering Center, Liquid Crystal Mie Group, Sharp Corporation, Tenri, Japan, June 6, 1997.

**TAKAAKI TANAKA,** chief researcher, LCD Development Center, LCD Operations Division, Seiko Epson Corporation, Suwa, June 13, 1997.

**STEPHEN A. TANSKI,** second secretary, environmental, scientific, and technological affairs, Embassy of the United States of America, Tokyo, Japan, October 18, 1996.

**REX TAPP,** president and CEO, Optical Imaging Systems, Inc. (OIS), Northville, Michigan, May 7, 1996.

**JEFF TETZLAFF,** vice president, sales and marketing, Display Technology Systems, St. Paul, Minnesota, September 11, 1996.

**MALCOLM THOMPSON,** chief executive officer, dpiX, a Xerox Company, and board chair, U.S. Display Consortium (USDC), September 12, 1996, Boston, Massachusetts; Palo Alto, California, October 22, 1998.

**ARTHUR Q. THOMPSON,** division engineering manager, Advanced Display Products, Corning, Incorporated, September 13, 1996.

**CHUN-HUI TSAI,** manager and project leader, Vacuum Microelectronics Development Department, Industrial Technology Research Institute, Electronics Research and Service Organization (ITRI/ERSO), Hsinchu, Taiwan, May 28, 1997.

**HORACE TSIANG,** chief executive officer, First International Computer, Inc., Taipei, Taiwan, May 26, 1997.

**ASAKO TSUDA,** Planning Department, TFT-LCD Development Center, Liquid Crystal Display Group, Sharp Corporation, Tenri, Japan, June 6, 1997.

**HSING C. TUAN,** president, Unipac Optoelectronics Corporation, Hsinchu, Taiwan, May 29, 1997.

**M.J. TZOU,** vice president, Electronic Material Division, Nan Ya Plastics Corporation, Taipei, Taiwan, May 26, 1997.

**DENNOSUKE UCHIDA,** president, Japan, Semiconductor Equipment and Materials International (SEMI), Tokyo, Japan, October 21, 1996.

**TOSHIHIRO UEKI,** general manager, Engineering #2, Display Technologies, Inc. (DTI), Himeji, Japan, June 2, 1997.

**RICHARD VAN ATTA,** assistant deputy under-secretary for dual use and commercial programs, United States Department of Defense, Washington, D.C., December 19, 1995.

**J. GERARD (JERRY) VIEIRA,** director of strategic marketing, Planar Systems, Inc., Beaverton, Oregon, June 10, 1996.

**M. KATHY VIETH,** president/COO, Coloray Division of Scriptel Holdings, Inc., San Jose, California, July 2, 1996.

**SHINJI WADA,** project leader, LCD Development Center, LCD Operations Division, Seiko Epson Corporation, Suwa, Japan, June 13, 1996.

**LARRY F. WEBER,** president, Plasmaco, Highland, New York, July 24, 1996.

**MAXWELL S.C. WEI,** unit manager, LCD Department, Electronic Materials Division, Nan Ya Plastics Corporation, Taipei, Taiwan, May 26, 1997.

**HAROLD J. WIENS,** executive vice president, Sumitomo 3M Limited, Tokyo, Japan, June 10, 1997.

**CHARLES C. WILSON,** executive vice president and chief financial officer, Optical Imaging Systems (OIS), Northville, Michigan, May 7, 1996.

**ROBERT WISNIEFF,** manager, flat panel display fabrication, International Business Machines Corporation (IBM), Thomas J. Watson Research Center, Yorktown Heights, New York, July 22, 1996.

**ANDREW WONG,** vice president, 3M Optical Systems Division, St. Paul, Minnesota, November 21, 1997.

**EIJI YAMAGUCHI,** manager, Development Dept. 12, FV Project, Div. 1. Canon, Inc. R&D Headquarters, Canon Research Center, Atsugi, Japan, November 5, 1996.

**ROBERT L. YARD,** marketing director, Advanced Display Products, Corning, Incorporated, New York, September 13, 1996; Tokyo, Japan, November 7, 1996.

**MASAAKI YASUKAWA,** manager, Multimedia Planning Center, Seiko Epson Corporation, Tokyo, Japan, November 8, 1996; Suwa, Japan, June 13, 1997.

**OON SEOK YEO,** deputy director, Electronic Components and Consumer Electronics Division, Ministry of Trade, Industry and Energy, Republic of Korea, Seoul, Korea, December 10, 1996.

**TERUO YONEYAMA,** senior manager, Liquid Crystal Display Division, Production Administration Department, Toshiba Corporation, Tokyo, Japan, November 5, 1996, and June 12, 1997.

**TODD YUZURIHA,** director, LCD technology R&D, Corporate Strategic Engineering Center, Sharp Microelectronics Technology, Inc., Camas, Washington, June 12, 1996.

# APPENDIX FOUR

**GLOSSARY OF ACRONYMS AND ABBREVIATIONS**

*(exclusive of company names)*

AC PDP: alternating current plasma display panel
ADMA: Advanced Display Manufacturers of America
AMLCD: active matrix liquid crystal display
CRT: cathode ray tube
DARPA: Defense Advanced Research Projects Administration (U.S.)
DoD: Department of Defense (U.S.)
DRAM: dynamic random access memory
EDIRAK: Electronic Display Industrial Research Association of Korea
EL: electroluminescent (display)
FPD: flat panel display
FED: field emission display
IC: integrated circuit
ITC: International Trade Commission (U.S.)
ITO: indium tin oxide
LC: liquid crystal
LCD: liquid crystal display
MITI: Ministry of International Trade and Industry (Japan)
MoST: Ministry of Science and Technology (Republic of Korea)
NFPDI: National Flat Panel Display Initiative
PC: personal computer
PDP: plasma display panel
PMLCD: passive matrix liquid crystal display
SID: Society for Information Display
STN LCD: super-twisted nematic liquid crystal display
TFT LCD: thin-film transistor liquid crystal display
TN LCD: twisted nematic liquid crystal display
TRP: Technology Reinvestment Program
USDC: U.S. Display Consortium

# NOTES

In these endnotes, we have used shortened titles to refer to books, academic articles, and presentations that appear with complete citations in the book's "References" section. Citations to periodicals, print news media and World Wide Web sources appear in their entirety in the endnotes. In cases where a particular source received multiple citations within a chapter, notes that follow the first complete citation refer the reader back to it. Interview citations, also provided in short form here, appear in complete form in Appendix 3, "Interviews."

## CHAPTER 1: INDUSTRY CREATION AS KNOWLEDGE CREATION

1. Both accessed via Lexis-Nexis, September 18, 2000.

2. Tracy Kidder's Pulitzer Prize-winning *The Soul of a New Machine,* first published in 1981, has achieved the status of a classic treatise on technology management and organization. The book told the story of a crash project at Data General to build a new 32-bit minicomputer in one year.

3. We estimated the number of fabs producing large-format, color amorphous silicon thin-film-transistor LCDs, with a processing capacity of at least 10,000 substrates per month, based on DisplaySearch, "Flat Panel Fab Database." When less-advanced FPD technologies such as TN and STN are included, the proportion of locations in Japan increases. We explain the different technologies in detail in Chapter 3.

4. We concern ourselves here with innovation processes that result in specific new products being offered in markets, rather than the basic research that may underlie a number of specific innovations.

5. W.J. Abernathy and J.M. Utterback. "Patterns of Industrial Innovation"; Giovanni Dosi, "Technological Paradigms"; David J. Teece, "Profiting from Technological Innovation." Teece draws attention to the parallel between the prior-named authors' perspectives on industry and Thomas Kuhn's conceptualization of scientific evolution in *The Structure of Scientific Revolutions.*

6. Hiroshi Take interview (see Appendix 3 for specific information on all interviews).

7. Webster E. Howard, "Foreword."

8. In Chapter 3 we will describe PDP, FED, and EL technologies, three alternative FPD approaches, each of which offers its own relative strengths for particular applications.

9. Toru Shima interview.

10. Yuko Inoue, "Market Slump Snags Color LCD Venture; Toshiba, IBM Revise Targets but Remain Optimistic," *Nikkei Weekly,* April 25, 1992, 8; "Display Tech to Market Large Color LCDs," Jiji Press Ticker Service, February 4, 1992, accessed via Lexis-Nexis September 1996; Masaki Higashi, "Electronics Firms Delay Investment in LCD Production, But Industry Giants Will Retain Their Leading Roles," *Nikkei Weekly,* July 25, 1992, 8.

11. James Watt, for example, built his 1769 steam engine as an incremental improvement in a long line of steam engine designs dating back to 1680 and earlier. Thomas Edison's incandescent light bulb (1879) improved on a number of designs put forward by other inventors between 1841 and 1878 (Diamond, *Guns, Germs and Steel,* 244–5). The general observation may have originated with George Basalla, *The Evolution of Technology.*

12. Consider as an example the relationship between corporate strategy and organization structure, now axiomatic to all managers, which Alfred D. Chandler (*Strategy and Structure*) distilled in the early 1960s from the late nineteenth- and early twentieth-century histories of Dupont, General Motors, Sears, and Standard Oil. These companies were dealing with fundamental alterations in the nature and scope of U.S. markets brought on by telecommunications and railways. Consider also Raymond Vernon's description in *Sovereignty at Bay* of the international product life cycle, which he drew in the mid-1960s from histories of how U.S. corporations expanded internationally in the global economic reconstruction that took place in the wake of World War II.

13. See W. Brian Arthur, *Increasing Returns and Path Dependence in the Economy,* particularly Chapter 1 for an overview.

14. See, for example, Michael Borrus and Jeffrey A. Hart, "Display's the Thing."

### CHAPTER 2: WHAT'S WRONG WITH THIS PICTURE?

1. "Long Term FPD Outlook: $70 billion in 2005," *DisplaySearch Monitor,* January 2000, 10–11.

2. Ross Young, "Current FPD Market Status."

3. Joseph A. Castellano, "Trends in the Global Flat Panel Display Industry." In fact, sales surpassed $18 billion in 1999. Estimates given in 1999 suggested that the 2001 market would exceed $28 billion. See Ross Young, "FPD Market Status and Outlook."

4. "Truly I say to you, no prophet is accepted in his own country." Luke 4:24. See also Mark 6:4 (Revised Standard Version).

5. Michael Porter originated the use of the word "cluster" to describe geographically concentrated groupings of interdependent economic activities in his book, *The Competitive Advantage of Nations.* See also his more recent article, "Clusters and the Economics of Competition."

6. For example, see important contributions by Richard R. Nelson, *National Innovation Systems,* and Bruce Kogut, *Country Competitiveness.*

7. The Nobel Prize-winning economist Kenneth J. Arrow referred to this dynamic in *Essays in the Theory of Risk Bearing* as the fundamental paradox of information.

8. United States Ambassador to Japan Walter F. Mondale interview.

9. Comment from the audience at a workshop given by Lenway and Murtha sponsored by the Asian Technology Information Program (ATIP), Tokyo, Japan (October 17, 1996).

10. United States Department of Defense, *Building U.S. Capabilities in Flat Panel Displays.*

11. The initiative was announced on April 26, 1994.

12. As recently as 1999, presentations were offered at an industry conference that divided the output of Display Technologies, Inc. (DTI) between its co-owners, Toshiba and IBM. DTI held the leading market share in the most advanced large-format displays in 1996 and 1997, and ranked either first or second in 1998. By issuing announcements based on a division of DTI's market share between its owners, competitors could assert that their companies held higher-ranking market shares than would have been the case had they reported DTI's market share in aggregate. Market researchers, as well, often apportion DTI's output between its owners, explaining that the assumptions underlying their reporting methodology attribute market share on a "shipment basis." Yet these same reports do not similarly apportion the output of the Korea-based LG.Philips LCD alliance between LG (headquartered in Korea) and Royal Philips (headquartered in the Netherlands).

13. The AKT alliance ended in 1998, according to Applied Materials' 1999 Annual Report, and Applied Materials has since successfully carried the business forward on its own as AKT, an Applied Materials Company.

14. Soichi Nagamatsu interview.

15. Ikujiro Nonaka and Hirotaka Takeuchi, *The Knowledge Creating Company,* 59.

16. For example, in early 1996, output plunged during the transition from second- to third-generation manufacturing lines. Companies that adopted third-generation lines could make six 12.1-inch displays from one glass substrate. Generation 2 lines could produce four 10.4-inch displays per glass substrate, but were converted to produce two 12.1-inch displays as consumer preferences shifted rapidly toward the larger displays. Overall industry capacity dropped significantly, creating shortages that were overlooked by many U.S. observers until a talk by Norihiko Naono given at the Asian Technology Information Program (ATIP) FPD Forum in Portland, Oregon, on June 20, 1996. Naono is a Rambus executive and former FPD industry analyst with Nomura. Makoto Sumita, a senior economist with the Center for Policy Research, Nomura Research Institute, also anticipated the shortage in remarks given as part of a February 5 panel discussion during the Business Conference that preceded the 1996 Display Works Annual Conference sponsored by the U.S. Display Consortium (USDC). But his remarks had little impact on many U.S. analysts, who anticipated that 10.4-inch screens would dominate supply through 1998, leading to a continuing glut.

17. For examples, see Nomura forecasts quoted in "Fast Forward: Survival of the Fittest," *The Economist,* UK edition, April 13, 1991, 14; 1992 Stanford Resources forecasts quoted in G. Christian Hill, "Motorola Video-Display Venture Aims to Wrest Market Share from Japanese," *Wall Street Journal,* Eastern Edition, October 22, 1992, B10; and 1993 Nomura estimates quoted in 1994 by Hager Grob, "The Flat Panel Display Industry in Japan," 5.

18. Larry Weber interview.

19. Lisa Ferrero, "AMLCD Substrate Market."

20. Tsuyoshi Kawanishi interview.

21. Steve Depp in conversation with Murtha and Lenway, Ypsilanti, Michigan, September 19, 1997.

22. Robert Wisnieff interview, July 22, 1996.

23. http://www.lucent.com/news/pubs/annual/96/pdf.first.pdf, accessed December 5, 2000.

24. Shinichi Hirano interview.

25. Sherwood Anderson, *Winesburg, Ohio.*

26. Edgar Lee Masters, *Spoon River Anthology.*

### CHAPTER 3: CONTINUITY UNDER ADVERSITY

1. Interview.

2. Revised Standard Version. See also Mark 6:4.

3. This account is based on an interview with Larry Weber. For another account, see Evan Ramstad, "A Passion for Plasma Helps Fuel Advances in Flat-Screen Technology: How Larry Weber Rescued Plasmaco, Seeks to Make High-Tech TVs Cheaper," *Wall Street Journal,* August 30, 2000, B1.

4. Plasmaco press release, May 15, 2000, accessed at www.plasmaco.com on March 28, 2001.

5. W. Brian Arthur, "Increasing Returns and the New World of Business."

6. Jared Diamond, *Guns, Germs and Steel;* George Basalla, *The Evolution of Technology.*

7. Competitive advantage in the initial phases of an industry lies elsewhere than in achieving technological superiority per se. For example, VHS triumphed in consumer markets over the Beta home videotape format, despite the latter's widely acknowledged technical advantage. See Richard S. Rosenbloom and Michael A. Cusumano, "Technological Pioneering: The Birth of the VCR Industry." We also know that in the English-speaking world, the "QWERTY" typing keyboard persists, despite its nineteenth-century origin as a device to *dis*advantage human users so they could not jumble the primitive mechanics of early typewriters by working too quickly. See Paul David, "Clio and the Economics of QWERTY." Though an element of chance clearly enters into the outcomes of technology races, stories like Larry Weber's and others recounted in this book suggest that building knowledge advantage amounts to something more than getting lucky.

The answers also lie elsewhere than in being in the right place at the right time, or enjoying the benefits of government industrial strategy. At the turn of the millennium, the impression remained widespread that Japanese companies enjoyed special (often described as "unfair") advantages in developing the FPD industry. Shortsightedness in U.S. companies and government, commentators argued, offered Japanese companies an early, probably insurmountable lead in commercializing FPD technologies. In FPDs, as in other industries, these critics suggested, Japanese companies seized the opportunity by gaining government assistance early in the industry's development and coordinating with each other to exclude competitors from other countries as the industry evolved.

We found no evidence of this. Furthermore, these arguments overlook the reality of Japan's economic struggle in the 1990s. The success of Japanese FPD companies and Japan's early dominance as a production site stood as an exception among Japanese industries during the 1990s.

Japanese economic growth fell into negative territory in thirteen out of thirty-eight quarters between 1991 and 2000. Nor can we fall back on a revisionist stance that attributes Japan's FPD industry role to superior Japanese management technique. Companies from Europe, the United States, Korea, and Taiwan all shared robust industry positions in the year 2001.

8. NEC started up a line in August 1990 that processed 300 by 350 mm substrates. The DTI and Sharp facilities respectively processed 300 by 400 and 320 by 400 mm substrates and started up about one year later. DTI and Sharp are sometimes credited with first movership in high-volume TFT LCD production because their approaches could yield either four 8.4-inch or two 10.4-inch displays, while the NEC approach at the time yielded two 9.4-inch displays and was regarded by some observers as a transitional step between Generation 0 and Generation 1. But NEC established an early lead in manufacturing yields. See "NEC to Up Color LCD Production," Jiji Press Ticker Service, November 14, 1991, accessed via Lexis-Nexis, March 31, 2001; and Hideki Wakabayashi, "Liquid Crystal Displays: The Next Five Years," 68 69, notably Table 7.

9. RCA was acquired by GE in 1986. In February 1987, the Sarnoff Center became a subsidiary of SRI International, an independent non-profit research institute with historical and board-level ties to Stanford University. In 2000, Sarnoff was operating as a for-profit subsidiary of SRI.

10. Tsuyoshi Kawanishi interview.

11. "RCA Develops a New Visual Display Means Using Liquid Crystal." *Wall Street Journal,* May 29, 1968, 4.

12. Bob Johnstone, *We Were Burning,* 96.

13. Ibid., 89. Johnstone's account conveys an optimism that may have been privately expressed by the researchers, particularly Heilmeier, whom he interviewed in March 1994 for his fascinating, valuable book. Most of the contemporary published journalistic accounts we accessed conveyed cautious or neutral assessments of the technology's likely time-to-market in television form. See for example, "Liquid Crystals," *Science Digest,* December 1968, 32–34. Yet journalists who were working in the field of consumer electronics around the time of the announcement and in the years following remember a sense of immediacy and excitement that contemporary journalistic style may have tempered in print. Telephone discussion, Robert Angus, former senior editor of *Consumer Electronics Monthly,* August 5, 2000.

14. Material in this paragraph relied on Joel Brinkley, *Defining Vision,* 51; Bob Johnstone, *We Were Burning,* 35, 102–105; and Linden Harrison, "Liquid Crystal Displays." We based the date of RCA's exit from LCDs on the company's 1968–75 Annual Reports. Mention of LC research and manufacturing disappeared from these reports in 1975.

15. Daniela Hager Grob, "The Flat Panel Display Industry in Japan."

16. Steven W. Depp and Webster E. Howard, "Flat-Panel Displays," 94.

17. Ibid.

18. Hiroshi Take, Asako Tsuda and Kotaro Kazuta interview.

19. Jon A. Shroyer and Joel Pollack interview.

20. According to Johnstone (*We Were Burning,* 38) an earlier fully electronic calculator by Sony, the SOBAX, never advanced beyond the prototype stage.

21. Kan Numakami, Ikujiro Nonaka, and Takebi Ohtsubo, "Sharp Technology Management," 4.

22. Asada was a young engineer of about five years' experience at Sharp when he was assigned to the LCD engineering team. As senior executive vice president at Sharp, he published a serialized memoir of his experiences and Sharp's development of LCDs in the *Asahi Shimbun* newspaper between July 5 and August 30, 1995. These memoirs, "The Changing World of LCD Displays," were provided to the authors in translation during a visit to Sharp's facilities in Camus, Washington, on June 12, 1996. This paragraph drew on the July 7, 1995, installment, "A Spooky Calculator is Born."

23. Quoted by James McGroddy, IBM senior vice president, research (retired), among others. Since 1989, Ralph Gomory has served as president of the Alfred P. Sloan Foundation, which funded the research that led to this book.

24. The international product cycle model ranks among the most durable observations encompassed within the late Raymond Vernon's immense, influential body of scholarship and policy analysis. His 1971 book *Sovereignty at Bay* foreshadowed the interaction of learning, technology, and globalization in industrial competition by at least twenty years. The book remains more relevant to understanding contemporary global competition than many later books by other authors. See also Vernon's most recent book, *In the Hurricane's Eye.*

25. Iridium officially ended service to customers March 17, 1999, at midnight, after receiving court permission to cease its search for a buyer. Some customers may have received erratic service in the months that followed. Steve Lohr, "Economic View: When You are First and Still Don't Succeed," *New York Times,* March 19, 2000, accessed via *New York Times on the Web,* August 12, 2000; and Michael A. Hiltzik, "Satellite Venture Will go Down in Flames, Literally," *Los Angeles Times,* March 18, 2000, A-1.

26. Press release at www.iridium.com, accessed April 3, 2001.

27. See, for example, Gary Hamel and Yves L. Doz, *Alliance Advantage.*

28. Material in the preceding three paragraphs relied on Daniela Hager Grob, "Flat Panel Displays in Japan," 10–14; Bob Johnstone, *We Were Burning,* 108–110, and on interviews at Seiko Epson's Tokyo offices on November 7, 1996, and in Suwa on June 13, 1997.

29. "Overview of New Sharp LCD Business," Sharp Corporate Document, Tenri, June 1997, contains a chronology of Sharp's LCD development.

30. In addition to the authors' interview notes from various company visits, this box relied on Steven W. Depp and Webster E. Howard, "Flat-Panel Displays," Daniela Hager Grob, "Flat Panel Displays in Japan," and Teruhiko Yamazaki, Hideaki Kawakami, and Hiroo Hori, eds., *Color TFT Liquid Crystal Displays.*

31. Hiroshi Take interview.

32. Bob Johnstone, *We Were Burning,* 128.

33. Sources differ in their reports of the size of the prototype. Damian Saccocio, "Strategy and New Business Development," 60, reports that the company demonstrated a one-inch diagonal prototype. Johnstone suggests that the prototype was two inches (*We Were Burning,* 118–120).

34. Several companies have pursued the electroluminescent (EL) approach to FPDs, including Sharp, Tektronix, and its spinoff, Planar, both U.S. companies. Electroluminescent displays produce light in much the same way that electrons produce light in the phosphors of a cathode ray tube. In EL manufacturing, a phosphor film, typically manganese-doped zinc sulfide, is deposited onto a glass substrate in a vacuum chamber. This phosphor is sandwiched

between two thin transparent insulating films, which are sandwiched between two thin transparent films that create a matrix of conductive row and column electrodes. When a voltage is applied to an individual pixel, the electrons that lie on the insulator-phosphor interface bombard the phosphor and create an amber image. EL displays are rugged, bright, have a long lifetime, and have a wide viewing angle. Their commercial success has been limited by their relatively high cost and inability to reproduce color. Sharp's early EL displays did, however, find application in the earliest laptop computer displays and were used in the original NASA Space Shuttle. In the late 1990s, EL applications were mostly confined to industrial applications that demand ruggedness.

35. T. Peter Brody, "The Birth and Early Childhood of Active Matrix." For a secondary account, see Richard Florida and David Browdy, "The Invention That Got Away."

36. Terry J. Scheffer, "The 'Iron Law' 25 Years Later." According to Scheffer, the original insight was Paul Alt's, but he found it obvious. His IBM supervisor Peter Pleshko, however, recognized the counterintuitive nature of Alt's result as well as its significance and persuaded him to publish it in a paper that became a classic, "Scanning Limitations of Liquid Crystal Displays," to which he also contributed.

37. Steven W. Depp and Webster E. Howard, "Flat-Panel Displays," 96.

38. In addition to the authors' interview notes from various company visits, this box relied on Steven W. Depp and Webster E. Howard, "Flat Panel Displays"; Daniela Hager Grob, "Flat Panel Displays in Japan"; and Teruhiko Yamazaki, Hideaki Kawakami, and Hiroo Hori, eds., *Color TFT Liquid Crystal Displays.*

39. Chronology presented in "Overview of New Sharp LCD Business," Sharp Corporate Document, Tenri, June 1997.

40. Hiroshi Take interview.

41. Mii eventually moved on to a senior vice presidency at IBM and a series of global responsibilities.

42. Bob Johnstone, *We Were Burning,* 56. According to Shinichi Hirano, collegial contacts also occurred regularly between IBM and Sharp researchers.

43. Tsuyoshi Kawanishi interview.

44. Damian Saccocio, "Strategy and New Business Development," 59–65.

45. Steve Depp, Robert Wisnieff, and (by teleconference) John Ritsko interview, July 22, 1996.

46. Saccocio, "Strategy and New Business Development," 60.

47. Larry Weber interview.

48. Donald Bitzer, "Inventing the AC Plasma Panel."

49. Steven W. Depp and Webster E. Howard, "Flat Panel Displays."

50. Alan Sobel, "Television's Bright New Technology," 72.

51. See www.ece.uiuc.edu/honors/bit.html, accessed March 28, 2001.

52. Howard had been pushing for TFT LCD development at IBM since the early 1980s, when he recognized the significance of research led by Walter Spear and Peter G. LeComber of the University of Dundee, Scotland. The research showed the suitability of amorphous silicon for fabricating TFTs. See Saccocio, "Strategy and New Business Development," 60. The company had operated research programs on a number of display technology alternatives to plasma at low levels of funding since the early 1970s. These included TFT LCDs using polycrystalline

silicon, an area in which Steve Depp's research made early contributions. In time, Howard won him over to amorphous silicon. Polycrystalline silicon posed difficulties as a potential material for high-volume, large-format TFT fabrication, because it required costly substrate materials such as quartz to withstand relatively high deposition temperatures. Amorphous silicon TFTs could be deposited at lower temperatures, making possible the use of glass substrates.

53. James McGroddy interview.

54. Field emission occurs when an externally applied electric field at the material surface thins the potential barrier to the point where electron tunneling occurs. Because no heat is involved, field emitters are referred to as a "cold cathode" electron source. Robert B. Smith, "Electronics Developments for Field Emission Displays."

55. The preceding paragraphs relied on interviews with James McGroddy; Steven Depp, Robert Wisnieff, and John Ritsko, July 22, 1996; and on Saccocio, "Strategy and New Business Development." Ironically, the IBM PC business continued to oppose the project and expressed reluctance to source TFT LCDs internally until new management took the reins, well after the first DTI factory was under construction.

56. James C. Morgan and J. Jeffrey Morgan, *Cracking the Japanese Market*, 100–103, 185.

57. Shinichi Hirano and Kohki Nohda interview.

58. James McGroddy interview.

59. This paragraph and the following section relies on the following interviews: Robert Yard and Gerry Fine, September 13, 1996; Yard, Lisa Ferrero, Chip Jones, Akihasa Mitsuhashi, and Satoshi Furuyama, November 7, 1996; Furuyama and Jones, June 16, 1997. Interviewees regarded many of the early dates mentioned in the interviews as approximate. The corporate website is similarly ambiguous. Wherever possible we have tried to triangulate among several interviews or third-party sources.

60. T. Peter Brody, "The Birth and Early Childhood of Active Matrix."

61. The collation of our chronology with James G. Kaiser's career was informed by Kevin D. Thompson's biographical article, "Blazing New Trails," *Black Enterprise,* January 1991, 54, accessed via Lexis-Nexis, August 22, 2000.

62. Chronology at http://advanceddisplay.corning.com, accessed August 22, 2000.

63. T. Peter Brody, "The Birth and Early Childhood of Active Matrix."

64. Atsushi Asada, "Liquid Crystal Alchemists," *Asahi Shimbun*, July 19, 1995.

65. Steven Depp interview, July 22, 1996. Retold in September 1997.

66. Tsuyoshi Numagami, "Strategic Deviation from the Trend of Technological Evolution," 28.

67. Hiroshi Take interview.

68. Out of sensitivity to U.S. antitrust law, IBM has since the 1950s remained reluctant to announce technology breakthroughs prior to the availability of products in markets.

69. "Sharp Develops Thin, 14-Inch TV Monitor Light Enough to Hang on Wall." Asahi News Service, June 17, 1988; "Toshiba, IBM Claim Largest Color Liquid Crystal Display," Japan Economic Newswire, September 21, 1988, accessed via Lexis-Nexis September 18, 2000.

70. Hiroshi Take interview; chronology presented in "Overview of New Sharp LCD Business," Sharp Corporate Document, Tenri, June 1997.

71. "Overview of New Sharp LCD Business," Sharp Corporate Document, Tenri, June 1997.

72. Hiroshi Take interview.

73. Webster E. Howard, "Foreword."


## CHAPTER 4: KNOWLEDGE AND COMMERCIALIZATION

1. Interview.

2. Interview.

3. O'Mara, William C. "Active Matrix Liquid Crystal Displays Part I: Manufacturing Process." *Solid State Technology,* December 1991, 65–70.

4. According to DisplaySearch, CVD absorbed 22 percent of the capital equipment budget for a new fab at the end of the 1990s. See Ross Young, "FPD Equipment and Materials Market Trends and Forecast."

5. In the years that have followed the visit described in this section, the authors continued to encounter industry participants and observers all over the world who insisted that AKT's manufacturing must take place in Japan.

6. Applied Materials announced the joint venture's creation on June 17, 1993. According to Applied's 1999 Annual Report, the venture ended in 1998. AKT then reorganized as a wholly owned subsidiary of Applied Materials. Since the reorganization, Iwasaki has served as AKT's chairman, in addition to his position as chairman and CEO of Applied Materials Japan.

7. Joseph Paone interview.

8. Webster E. Howard, "Foreword."

9. Hiroshi Take, Kotaro Kazuta, and Asako Tsuda interview.

10. Kanro Sato interview.

11. Kawanishi was executive vice president and head of Toshiba's worldwide electronics components and semiconductor businesses, and later held the title of senior executive vice president for partnerships and alliances. When the authors met with him on November 15, 1996, at Toshiba's Tokyo headquarters, he held the title of senior advisor and played a visible emeritus role within the company.

12. Canon and Nikon dominate the market for steppers. Anelva manufactures CVD equipment for the FPD industry, while Toray manufactures color filters.

13. James C. Morgan and J. Jeffrey Morgan, *Cracking the Japanese Market,* 123–143.

14. Norihiko Naono interview.

15. Steven W. Depp interview, July 22, 1996.

16. John Verity, "What's Ailing IBM? More Than This Year's Earnings," *BusinessWeek,* October 16, 1989, 75.

17. Damian Saccocio, "Strategy and New Business Development," 59–65.

18. Hiroshi Take interview.

19. In 1985, while investigating potential TFT LCD development partnerships, IBM's Barbara Grant and Steve Depp visited Sharp and were given a tour of an exhibition area in which the company showcased its most significant products. Noticing an empty pedestal, Depp inquired whether the space was reserved for the TFT LCD. The hosts confirmed, with some amusement, that it was indeed so. Telephone discussion, Steven Depp, November 14, 2000. The story of Sharp's "empty pedestal" has gained something of the status of an industry

legend, as the authors encountered some version of it in numerous conversations in the U.S., Japan, and Korea.

20. Hiroshi Take interview.

21. Steven W. Depp, Robert Wisnieff, and (by teleconference) John Ritsko interview, July 22, 1996.

22. "Toshiba, IBM Set Plant For Large LCDs." *Los Angeles Times,* August 30, 1989, Business Section, Part A, 3.

23. Establishing the precise historic capacities of substrate lines is extremely difficult because published data often quoted production figures rather than actual capacity and cell sizes varied, as did yields. The earliest large-format TFT fabs achieved yields of less than 10 percent at startup. We established the range of our estimate based on a September 1992 announcement that Sharp would increase production capacity on its original TFT LCD production line from 45,000 to 90,000 8.4-inch displays per month, cutting four from each 320 by 400 mm substrate (Japan Economic Newswire, September 21, 1992, "Sharp to Beef up LCD Capacity," accessed via Lexis-Nexis, September 1996). Initial capacity had been announced as 10,000 units per month (Asahi News Service, September 9, 1991, accessed via Lexis-Nexis, September 1996). DTI's Line 1 production was reported at startup as 10,000 10.4-inch units, produced two-up on 300 by 400 mm substrates (Jonathan West and H. Kent Bowen, "Display Technologies, Inc.," 1), 20,000 substrates per month in 1993 (Brooke Crothers, "Sharp Sticks to Displays With High Yield Record," *Electronic News,* April 12, 1993, 15), and was ramped to 30,000 substrates per month sometime before the publication of the DisplaySearch marketing consulting company's June 1997 "Flat Panel Fab Database."

24. Robert Wisnieff interview.

25. The metaphor is drawn from Norman Maclean's *Young Men and Fire,* which recounts the 1949 Mann Gulch Fire disaster, in which thirteen firefighters lost their lives in a matter of minutes. Karl Weick has reflected in a number of articles on organizational principles to be drawn from Mann Gulch and Maclean's account. See, for example, "Prepare Your Organization to Fight Fires." Our reading of Maclean is, however, independent of Weick's.

## CHAPTER 5: STARTING FROM NOTHING

1. "Toshiba, IBM Japan Link to Make LCDs," *Nihon Keizai Shimbun (Japan Economic Journal),* September 9, 1989, 13.

2. "Sharp Up for Volume Color LCD Production," Jiji Press Ticker Service, October 12, 1989, accessed via Lexis-Nexis, September 18, 2000.

3. Yuko Inoue, "Market Slump Snags Color LCD Venture; Toshiba, IBM Revise Targets but Remain Optimistic," *Nikkei Weekly,* April 25, 1992, 8.

4. Ibid.

5. "Toshiba-IBM LCD Venture Firm Goes On-Line," Japan Economic Newswire, Kyodo News Service, May 15, 1991, accessed via Lexis-Nexis, September 26, 2000.

6. Jonathan West and H. Kent Bowen, "Display Technologies, Inc.," 6.

7. Steven Butler, "The Art of Perfection: Steven Butler Explains How a Single Speck of Dust Creates Havoc with the Way Liquid Crystal Displays are Made," *The Financial Times,* London, April 16, 1992, Technology Section, 14.

8. Bill Snyder, "Full Speed Ahead: Toshiba Corp. Plans Ambitious Expansion," *PC Week,* November 7, 1994, A1.

9. More specifically, Joji Itoh and Seiichiro Yokoyama, "Structure of Color TFT Liquid Crystal Displays," 149–279.

10. Steven W. Depp and Webster E. Howard, "Flat-Panel Displays," 94.

11. Joji Itoh and Seiichiro Yokoyama detailed this catalogue of steps in "Structure of Liquid Crystal Displays," 172–73.

12. According to Hiroshi Take, Sharp's young engineers discovered the use of ITO for this purpose as they waited impatiently for upper management's decision to proceed with high-volume production.

13. Steppers execute repeated exposures while sequentially changing small masks, called reticles, precisely exposing substrates in distinct sections. The distinct sections may pertain to the TFT arrays for multiple cells to be fabricated from one substrate, or to different (non-overlapping) sections of the same TFT array, depending on the intended size of the end product.

14. Photoresist comes in both positive and negative formulations. Positive-resist breaks down when exposed to light; negative-resist polymerizes or hardens when exposed.

15. The alignment film orients the thin-film molecules once the cell is filled (see the box in chapter 3, *Twisted and Supertwisted Nematic Liquid Crystal Displays*).

16. Temperature variation of as little as one degree in the exposure step can cause a substrate to expand or contract by one micron, creating the potential for misalignment of the mask and substrate. We have amply described the critical importance of maintaining cleanliness to exclude particles. Miniscule residues left over from the various coating and etching processes in the deposition phases may also corrupt subsequent chemical processes and materials, leading to malfunction.

17. The numbers come from Steven Butler, "The Art of Perfection," op. cit., note 7.

18. Yuko Inoue, "Production Woes Stall Mass-Market Hope for Color LCDs," *Nikkei* op.cit., note 3. The article reported that Nomura analysts had identified color filters and CVD efficiency as "two major headaches for LCD parts makers, partly due to limited competition." Color filters were expensive and in short supply, and CVD inefficient, causing production bottlenecks.

19. Brooke Crothers and Jack Robertson, "IBM/Toshiba LCD Unit Sees Yields Rise to 50 Percent," *Electronic News,* July 6, 1992, accessed via Lexis-Nexis, July 1999.

20. Yuko Inoue, "Production Woes Stall Mass-Market Hope," op. cit., note 3.

21. Ibid.

22. Steven Butler, "The Art of Perfection," op. cit., note 7.

23. Toru Shima interview.

24. Steven Butler, "The Art of Perfection," op. cit., note 7. Ironically, five years after Satoh made his humorous comment to Butler about the dust generated by operators sitting down in the fab, the authors of this book witnessed a cleanroom operator in action who did not sit down,

but instead sat a cassette of substrates precariously on an office chair to roll it from one production station to another. "Slight breakdown of cleanroom discipline," commented the company president, who was leading us on a tour. Two years later, the company left the industry.

25. Companies tend to be secretive concerning their precise yields. Many conflicting claims for different time periods appeared in the media during this time. Claims in our interviews also tended to conflict with other records. We have relied on figures that appear generally consistent across multiple sources. These tend to fall considerably short of the most flamboyant claims companies made about themselves and somewhat beyond the claims companies sometimes made about others. The NEC and DTI figures appeared in Steven Butler, "The Art of Perfection (op. cit., note 7)." Jonathan West and H. Kent Bowen stated in "Display Technology, Inc." that DTI yields were 44.6 percent some time in 1993, and built upward from there. In "Update on Flat Panel Displays," the respected LCD and semiconductor materials and equipment technical marketing consultant William O'Mara provided the 50 percent industrywide figure for 1993 as well as the DTI estimate. The authors could not verify the NEC claims from any alternative source. Informants typically cautioned us against directly comparing companies' yield claims, as different firms were reputed to use different measures, or to offer only the yields of particular production stages. Yields may also vary based on flow of capacity. As a senior manager at one manufacturer asked, "If the yield is 90 percent on 10 percent flow of capacity, what does this mean?" Generally, yields may be calculated as the product of yields from the different production stages. Using the three stages mentioned in the box in Chapter 5, *Manufacturing TFT LCDs,* a company that achieved apparently high yields of 70 percent in the deposition stage, 80 percent in the cell assembly stage, and 90 percent in the packaging stage would report a yield of just over 50 percent.

26. Toru Shima interview.

27. Jonathan West and H. Kent Bowen, "Display Technology, Inc."

28. Toru Shima interview.

29. John Ritsko interview.

30. This hope was realized by 1996.

31. Kanro Sato interview.

32. We acknowledge, as the basis for our observation, Ikujiro Nonaka and Hirotaka Takeuchi's discussion of the importance in knowledge creation of *Ba,* which they translated from Japanese as place. The Japanese concept of *Ba,* as reflected in Nonaka and Takeuchi's discussion in *The Knowledge Creating Company,* is multidimensional and not restricted to a physical location or even mental space in the limited sense that we have used it here.

33. Jonathan West and H. Kent Bowen, "Display Technology, Inc." The "swear box" is a commonplace, informal institution whereby co-workers agree to contribute a nominal sum as restitution for using words or language generally considered offensive in polite society. The group donates the contents of the swear box—which may be a jar, cigar box, or perhaps child's savings bank—to charity or may use it to partially fund some employee event. While the authors have not conducted a formal survey of the average magnitude of fines levied in swear box regimes, it seems fair to suggest that the DTI name-dropping fines (around $10) were quite stiff by comparison.

34. The quote is from Toru Shima. As of June 1997, less than one-third of DTI employees originated at either parent firm.

35. In February 1992, Toshiba and IBM announced that even with output of about 10,000 displays per month, they had accumulated surplus inventories and had decided to sell DTI's TFT LCDs outside the alliance within the year. "Display Tech to Market Large Color LCDs," Jiji Press Ticker Service, February 4, 1992, accessed via Lexis-Nexis, September 26, 2000.

36. Masaki Higashi, "Electronics Firms Delay Investment in LCD production, But Industry Giants Will Retain Their Leading Roles," *Nikkei Weekly,* July 25, 1992, 8.

37. Difficulties with particles generated by early CVD processes and slow CVD throughput were widely cited in our interviews as a major challenge to overcome in the transition from first- to second-generation production processes. See also Yuko Inoue, "Production Woes Stall Mass-Market Hope for Color LCDs," op. cit., note 3.

38. Tetsuo Iwasaki interview, October 24, 1996.

39. Zenzo Tajima interview, June 9, 1997.

40. At the time of this writing, Applied Materials' president.

41. Dennis Hunter interview.

42. Tetsuo Iwasaki interview, October 24, 1996.

43. Ibid.

44. For details, see Andrew Polack, "Applied Materials Plans Venture with Komatsu," *New York Times,* June 18, 1993, Section D, 3.

45. Tetsuo Iwasaki interview, October 24, 1996.

46. Material for this paragraph was taken from DisplaySearch, *Equipment and Materials Analysis and Forecast.*

## CHAPTER 6: SURVIVING THE KILLER APP

1. Tetsuo Iwasaki interview, October 24, 1996; Shinichi Hirano interview.

2. Hiroshi Take interview.

3. Peter Coy, "As Laptop Fever Spreads, IBM Trots Out a Heavyweight," *BusinessWeek,* November 26, 1990, 126.

4. Brooke Crothers, "Improved Color Displays Aim at Pricing Leverage," *Electronic News,* April 12, 1993, 13. Monochrome STN LCD manufacturing lines could be modified to produce color by adding a color filter to one of the substrates. In 1993, color STN LCDs cost about one half of the cost of color TFT LCDs.

5. Adam Greenberg, Brooke Crothers, and Jonathan Cassell, "Notebook Shortage Blamed on LCDs," *Electronic News,* December 14, 1992, 1.

6. Geoff Lewis, "IBM Strikes Back with a Laptop and Lower Prices," *Business Week,* April 14, 1986, 39.

7. Peter H. Lewis, "At IBM, Staying Ahead of the Bureaucracy Police," *New York Times,* March 14, 1993, Sunday Section 3, 9.

8. Speech by Robert Corrigan at COMDEX Show, Las Vegas, Nevada, November 16–20, 1992, excerpted as "IBM PC Co.'s Weight Reduction Plan: Start-up Sheds Corporate Layers," *Computer Reseller News,* December 7, 1992, 103. The idea of a separate personal computer company within IBM revived and formalized a heritage of independence in IBM's PC business. The original IBM PC was designed and brought to market in six months by a group set up as

a small business away from IBM's centers of operation and freed of IBM policies and practices. Over the years, IBM had gradually reabsorbed the business into its traditional structure.

9. John Schneidawind, "IBM Creates Subsidiary for Personal Computers," *USA Today*, Sept. 3, 1992, 3B.

10. Kristina Sullivan, "Color Notebooks Also Hit by PC Price Wars," *PC Week*, November 16, 1992, accessed via Lexis-Nexis, August 15, 2000.

11. "Color Display, Built-in Pointing Device Lead IBM ThinkPad Line," PR Newswire, October 5, 1992, accessed via Nexis-Lexis, August 15, 2000. One month after the product announcement, the editors of *PC/Computing* presented IBM with the "Most Valuable Product" (MVP) award at the annual "Best Hardware and Software of the Year" award ceremonies in Las Vegas before the opening of COMDEX Fall 1992. "IBM ThinkPad 700C Named Best Color Notebook Computer," PR Newswire, November 16, 1992, accessed via Lexis-Nexis, August 1999.

12. "IBM PC Company Surpasses 100,000 Orders for ThinkPad Notebook Computers," Business Wire, December 22, 1992, accessed via Lexis-Nexis, August 18, 2000.

13. Adam Greenberg, Brooke Crothers, and Jonathan Cassell, "Notebook Shortage," op. cit., note 5.

14. Brooke Crothers, "Sharp Sticks to Displays with High Yield Record," *Electronic News*, April 12, 1993, 15. The source quoted regarding the IBM PCC overture to Sharp was Nomura Research Institute's Norihiko Naono.

15. "Toshiba-IBM Japan Venture to Boost TFT LCD Output," Japan Economic Newswire, Kyodo News Service, July 8, 1993, accessed via Lexis-Nexis, September 26, 2000.

16. ED Research of Japan, *Full Details of Japanese Liquid Crystal Manufacturing Lines*, 7.

17. Ellis Booker, "Plethora of Panels," *Computerworld*, July 14, 1993, 33.

18. ED Research of Japan, *Full Details*, op. cit., note 16.

19. Michael Fitzgerald, "Manufacturers Vow End to Color Notebook Backlog by '95," *Computerworld*, August 29, 1994, 4.

20. Jack Robertson, "US. Plan Scotched—IBM/Toshiba Venture to Expand in Japan," *Electronic Buyers' News*, April 11, 1994, 1.

21. David Lammers, "LCD, DRAM Plants Planned," *Electronic Engineering Times*, July 11, 1994, 4.

22. Matsuada Hisajuki, "Advanced LCD Makers Look Beyond PCs: New Applications Sought as Production Ramp-up Brings Down Prices, Raises Worries of Oversupply," *Nikkei Weekly*, March 20,1995, accessed via Lexis-Nexis, July 2000.

23. Jacob M. Schlesinger, "New Computer Screen Plant Shows Production Can Snag in Japan, Too," *Wall Street Journal*, April 13, 1992, 1.

24. Norihiko Naono interview.

25. Michael Davis, "Quake Could Have Effect on Compaq," *Houston Post*, January 20, 1995, accessed via Lexis-Nexis, July 2000.

26. Kivka Tadjer, "A Kobe Aftershock Hits Laptop Industry," *New York Times*, January 29, 1995, 6.

27. Tom McHale, "Samsung TFT Biz to Grow," *Electronic Buyers' News*, February 20, 1995, 48.

28. Charles L. Cohen, "New LCD Screens Grow Bigger and Cost Less," *Electronic Buyers' News*, February 13, 1995, 8.

29. Jack Robertson, "LCD Industry Faces Shake-Up," *Electronic Buyers' News*, April 17, 1995, 3.

30. Jeff Bliss, "Notebooks," *Computer Reseller News*, June 3, 1996, 126.

31. Anthony Cataldo, "Display Buyers Rejoice," *Electronic Buyers' News*, December 18, 1995, 1.

32. "Koreans Push LCD Prices Lower," *Electronic Buyers' News*, August 28, 1995, 25; Mark LaPedus, "AM-LCD Price War Surges as Tags Fall 47 Percent in Taiwan," *Nikkei Weekly*, October 30, 1995, 2. The quote is from LaPedus.

33. William C. O'Mara, "How to Make Money in the Display Business."

34. David Lammers, "Cheap Color LCDs Cause Concern," *Electronic Engineering Times*, December 4, 1995, 24.

35. Brian Fuller, "Flat-Panel Glut Sparks Price Wars," *Electronic Engineering Times*, August 28, 1995, 1.

36. Anthony Cataldo, "Display Buyers Rejoice," op. cit., note 31.

37. Rick Knox, "User Views of the FPD Industry."

38. Norihiko Naono interview.

39. Pete Townshend, *My Generation*, (London: Fabulous Music, 1965).

40. Tetsuo Iwasaki interview, October 26, 1996.

41. Steven W. Depp, "Presentation."

42. In July 1996, the AKT 3500 received the grand prize for display manufacturing equipment at the Fine Process Technology Exhibition and Seminar in Tokyo. See "Applied Komatsu Technology Passes 75th TFT-System Shipment: Leadership in Flat Panel Display Equipment Market Supported by Multi-Chamber Architecture Patent," Business Wire, September 16, 1996, accessed via Lexis-Nexis, August 15, 2000.

43. AKT left the dry etch and sputtering markets in 1998 following the cancellation of a large order from a Korean TFT manufacturer after the equipment had already been built. The 1997 Asian Economic Crisis resulted in an unsustainable revenue decline for the company. AKT had entered the dry etch market late and had suffered from higher prices than its competitors. AKT also entered sputtering late and held only an 8-percent market share in 1998. DisplaySearch Email news bulletin, October 13, 1998.

44. XGA or fully extended graphics array compatible displays consist of 1024 by 768 pixels, while super video graphics array or SVGA have 800 by 600 pixels. Display resolution increases with the number of pixels.

45. DTI fab tour, Yasu, Japan, June 2, 1997.

46. Toru Shima gave DTI employment as 1,451 in June 1997.

47. Nikkei Business Press, *Nikkei Microdevices Flat Panel Display 1996 Yearbook*.

48. Jack Robertson, "FPD Players Split Over Glass Mfg. Strategy—Can't Decide On Shift To Larger Screens," *Electronic Buyers' News*, May 20, 1996, accessed via Lexis-Nexis, September 26, 2000.

49. Material on DTI in this section relied relied on Toru Shima, Masao Amano, Hiroshi Kimura, Masaru Shiozaki, and Toshihiro Ueki interviews.

50. Jack Robertson, "Suppliers Phase Out 10.4" Displays," *Electronic Buyers' News,* April 22, 1996, 1.

51. "NEC's New 12.1-inch LCD Display is Largest Available for Multimedia Notebooks," Business Wire, November 13, 1995, accessed via Lexis-Nexis, July/August 2000.

52. Hidenori Ikeda interview.

53. Itsuro Adachi interview.

54. Keinosuke Kamada interview.

55. In Chapter 4, we discussed Morozumi's role in the revolutionary invention and com-mercialization of color TFT LCD TV at Seiko.

56. "Japan's Hosiden to Produce LCD Monitors with Philips," *Asia Pulse,* March 11, 1997, accessed via Lexis-Nexis, July/August 2000; Keinosuke Kamada and Yasuki Iwano interviews.

57. Yoshiko Hara, "Philips Merger Spells Stability for Hosiden," *Electronic Engineering Times,* November 11, 1996, accessed via Lexis-Nexis, July/August 2000.

58. Richard L. Hudson, "Philips Refits Dutch Plant in Bold Plan to Unseat Rivals—Company to be Sole Challenger to Japanese Flat Panel Display Industry," *Wall Street Journal,* July 28, 1993. The partners were Merck and Sagen Thomsen.

59. J.C. Stuve interview.

60. Takahiso Hashimoto interview.

61. Steven Depp, Robert Wisnieff, and John Ritsko interview, July 22, 1996.

62. *DisplaySearch Monitor,* September 1997, p.1.

63. Jack Robertson, "Suppliers Phase Out 10.4-inch Displays," op. cit., note 50.

64. Ibid.

65. *DisplaySearch Monitor,* September 1997, p.1.

66. Junko Yoshida and Yoshiko Hara, "Japan's Top Companies Display New Ideas for Digital TVs, PDAs and Flat Panels at Show," *Electronic Engineering Times,* October 12, 1998, accessed via Lexis-Nexis, December 20, 2000; Janet Pinkerton, "Sharp Rides LCD Evolution," *Dealerscope,* December 1998, 24.

### CHAPTER 7: KNOWLEDGE-DRIVEN COMPETITION

1. See Alfred D. Chandler, *Scale and Scope.*

2. Kam Law, "Flat Panel Display Manufacturing Challenges."

3. This represents a conservative interpretation. The outcomes of such analyses depend critically on how many generations the analyst judges to have passed. This is in part an issue of industry politics, as companies may claim territory by referring to their own incremental changes as generation changes or by discounting the generational claims of others.

4. The work of Michael Polanyi is seminal and authoritative in this arena. See, for exam-ple, *Personal Knowledge* and *The Tacit Dimension.*

5. Ikujiro Nonaka and Hirotaka Takeuchi, *The Knowledge Creating Company,* especially Chapters 3 and 4.

6. See Karl Weick, *Sensemaking in Organizations,* particularly 88. Here Weick details a controversy among organizational researchers concerning how organization members deal best

with frequent and unpredictable directions of change, or turbulence. All partisans to the debate agree that as turbulence increases, people rely on less comprehensive information processes, such as intuition and heuristics, for understanding phenomena and making decisions. Innovation processes rely heavily on intersubjective sensemaking (72), which is to say that creativity thrives on direct interpersonal contact. Creative processes come under stress when participants begin to resort to shorthand or routine communications such as email (73).

7. Jae Choon Lim interview.

8. The papers of W. Bryan Arthur, of which many important ones are gathered together in the volume *Increasing Returns and Path Dependence in the Economy,* make this point particularly with respect to the establishment of industry standards, geographic concentration of an industry, information contagion, and many other important business phenomena. Arthur also examined the importance of his theories for business strategy in an article in the *Harvard Business Review,* "Increasing Returns and the New World of Business." Arthur's work relies on general principles now increasingly well-known under the aegis of the multidisciplinary field of complexity theory. M. Mitchell Waldrop has compellingly recounted Arthur's intellectual odyssey, and those of like-minded scientists in fields such as physics, biology, and computer science in his very accessible book *Complexity.* Shona L. Brown and Kathleen M. Eisenhardt have built on the basic science in the field to create and illustrate a set of management principles in their book, *Competing on the Edge.*

9. Our argument can be stated in economic terms. In these terms we would argue that the FPD industry's concentration in Japan emerged as a consequence of agglomeration economies with positive externalities. We believe that increasing returns to knowledge creation existed at the industry level in Japan, and that individual companies could maximize their shares in these benefits only by locating in Japan. It is important to note that we do not argue that the concentration of the FPD industry in Japan was inevitable. Japanese and U.S. firms made different decisions early in the industry's evolution. Most U.S. firms made discontinuous progress; several started and halted FPD development several times. Early Japanese adherents maintained steady commitments and created a continuing stream of increasingly sophisticated products that incorporated FPDs. Two relevant papers appear in Arthur's *Increasing Returns and Path Dependence in the Economy.* These are "Industry Location Patterns and the Importance of History" and "Information Contagion" (with David A. Lane).

10. In *The Knowledge Creating Company,* Ikujiro Nonaka and Hirotaka Takeuchi represented knowledge creation as a spiral. We have freely adapted their graphic approach which they designed to address individual, group, and company level processes, to illustrate our ideas about an industry-level phenomenon.

11. The re-examination of these assumptions has been urged since the late 1970s in business school dissertations and later in books that formed cornerstones of the international strategic management canon. We refer in particular to C.K. Prahalad and Yves L. Doz, *Multinational Mission,* and Christopher A. Bartlett and Sumantra Ghoshal, *Managing Across Borders.* The ideas presented in these books have found many ideological adherents among senior managers, but few practitioners until recently. Globalization, for many companies, has translated into centralization.

12. Michael E. Porter's "Competition in Global Industries" and Bruce Kogut's "Designing Global Strategies" remain seminal statements. Michael Y. Yoshino and U. Srinivasa Rangan, *Strategic Alliances,* provide a translation of value chain logic for interfirm collaboration.

13. Gary Hamel and C.K. Prahalad, *Competing for the Future,* 1994.

14. See Michael E. Porter, *Competitive Advantage,* and Bruce Kogut, "Designing Global Strategies," for comprehensive discussions of generic strategies.

15. Hamel and Prahalad, *Competing for the Future,* 1994.

16. See Richard R. Nelson, *National Innovation Systems;* also, Walter Kuemmerle, "Building Effective R&D Capabilities Abroad."

17. Many cost-driven companies retain responsibility for process design and improvements at home because they count these as core knowledge assets. Process design and improvement play critical roles in driving down costs. This is also true for companies pursuing differentiation strategies in industries such as chemicals, semiconductors, pharmaceuticals, and flat panel displays. In these industries, process design and improvements help to erect barriers against imitation of product-embedded advantages by affecting product performance and attributes.

18. Affiliates may play even more central roles in creating the knowledge basis of competitive advantages in service industries, particularly if we define services as Robert Stern did in a luncheon address that one of the authors attended in the late 1980s. Stern defined services as "goods for which manufacturing and consumption take place simultaneously."

19. Andrew Wong interview.

20. For evidence of this see Raymond Vernon, *In the Hurricane's Eye,* particularly 13, Table 1.6.

21. The academic literature supporting this orientation is vast. See Richard Caves, *The Multinational Corporation and Economic Analysis,* which is a book-length literature review.

22. See Dong Ho Lee, *Foreign Market Entry.*

23. As James Curry and Martin Kenney pointed out in "Beating the Clock: Corporate Responses to Rapid Change in the PC Industry," Dell's no-inventory approach and the "float" afforded by the time elapsed between the moment customers place prepaid orders and the moment that the company ships their computers offers a significant cost advantage in financing operations.

## CHAPTER 8: KNOWLEDGE AND NATIONALITY

1. Interview.

2. Interview.

3. Interview.

4. Stefanie Ann Lenway and Thomas P. Murtha, "The State as Strategist"; Thomas P. Murtha and Stefanie Ann Lenway, "Country Capabilities and the Strategic State."

5. U.S. Congress, Office of Technology Assessment, *The Big Picture,* 70.

6. U.S.-based production using other FPD technologies also stood at negligible levels. The 1994 U.S. domestic figure was first reported by Keith Bradsher, "U.S. to Aid Industry in

Computer Battle With the Japanese," *New York Times,* April 27, 1994, 1. The authors reviewed a printed version of the U.S. Department of Defense report from which this and the Japanese market share estimate was drawn, "Building U.S. Capabilities in Flat Panel Displays."

7. Authors' calculation. *DisplaySearch Monitor,* March 2000, p. 11, provided market share data for Samsung and LG.Philips. For Hyundai market share, see also Ross Young, "FPD Market Status and Outlook."

8. See Thomas P. Murtha, "Credible Enticements"; Thomas P. Murtha, "Surviving Industrial Targeting"; and Thomas P. Murtha and Stefanie Ann Lenway, "Country Capabilities and the Strategic State."

9. "Flamm Quits FPD Post," *Thin Film/Diamond Technology News,* August 1995, accessed via Lexis-Nexis, September 1995.

10. Thomas Kalil interview; former senior Department of Defense official interview by Murtha, Lenway, and Hart, 1995.

11. Correspondence, former U.S. Department of Defense official to authors, November 30, 1998.

12. Written communication, former U.S. Department of Defense official to authors, July 31, 1998.

13. Jay Stowsky, "America's Technical Fix"; Rose Marie Ham and David C. Mowery, "Enduring Dilemmas in U.S. Technology Policy."

14. Written communication, former U.S. Department of Defense official to authors, July 31, 1998.

15. Interview, industry official with Murtha and Lenway, Silicon Valley, summer 1996.

16. We adopt Stephen J. Kobrin's widely accepted definition of political risk as the risk of a change in public policies relevant to business decision-making. See *Managing Political Risk Assessment,* especially Chapter 3.

17. John Seely Brown and Paul Duguid, *The Social Life of Information,* 138–139.

18. Former U.S. Department of Defense official interview by Murtha, Lenway, and Hart, 1995; Jae-Choon Lim interview.

19. Jack Robertson, "US FPD Group: Koreans Go Home," *Electronic Buyers' News,* October 24, 1994, 1.

20. Written communication, former U.S. Department of Defense official to authors, July 31, 1998.

21. Griffith L. Resor III interview.

22. Jack Robertson, "MRS, Unlucky in Flat Panels, Shifts to PCBs," *Electronic Buyers' News,* April 6, 1998, 8.

23. Jack Robertson, "MRS Technology President Resigns," *Electronic Buyers' News,* March 2, 1998, 6.

24. Scott Holmberg interview. See also Jack Robertson, "Toshiba Turns to Hyundai for Flat Panel Aid," *Electronic Buyers' News,* November 4, 1996, 8.

25. At the conclusion of the Uruguay round of GATT negotiations in 1994, the institution became known as the World Trade Organization (WTO).

26. U.S. Defense needs, for example, were estimated to require no more than 15,000 FPDs per year at the time of peak military hardware retrofitting from 1995–1999. From

2000–2009, requirements were expected to reach 25,000 displays per year, and rise to perhaps 90,000 per year between 2010–2019. High-volume TFT LCD plants produce upwards of one million displays per year. Defense needs estimates appeared in U.S. Department of Defense, "Building U.S. Capabilities in Flat Panel Displays."

27. Yves L. Doz, Jose Santos, and Peter Williamson characterized this paradox for knowledge-based competition and treated it in their book *From Global to Metanational.*

28. "Planar Advance and Xerox Corporation Announce Alliance to Supply Displays to the Defense Market," Business Wire, March 13, 1995, accessed via Lexis-Nexis, November 25, 2000; Malcolm Thompson interview, October 22, 1998.

29. "Samsung to Exhibit Industry's Highest Quality Flat Panel Displays at COMDEX," Business Wire, November 8, 1994; Yoshiko Hara, "Korean Vendors Mount LCD Challenge to Japanese," *EETimes,* November 10, 1997, both accessed via Lexis-Nexis, October 1996.

30. Rex Tapp interview; Malcolm Thompson interview, October 22, 1998.

31. Rex Tapp interview; "The Days of OIS," *The Clock,* December 1998, 5–8.

32. "Korean TFT LCD Producers Expected to Match DRAM Share by End of the Decade," PR Newswire, August 27, 1997, accessed via Lexis-Nexis, July/August 2000.

33. Jun H. Souk interview.

34. Ibid.

35. "Samsung Breaks Ground on Third TFT LCD Production Line," Business Wire, January 7, 1997, accessed via Lexis-Nexis, July/August 2000.

36. Jun H. Souk interview.

37. Jerry Ascierto, "Death of a Dream? U.S. Hopes Fading in FPDs," *Electronic News,* April 5, 1999, 1.

38. Rex Tapp interview.

39. U.S. Department of Defense, "Building U.S. Capabilities in Flat Panel Displays," VIII-2–VIII-4.

40. Ibid., VIII-6.

41. "Special Defense Department Briefing, re: High-Tech Imaging Research," by Ken Flamm, principal deputy assistant secretary of defense, dual use technology and international programs, Federal News Service, April 28, 1994, accessed via Lexis-Nexis, February 1995.

42. M. Mitchell Waldrop. *Complexity,* 66.

43. Robert Duboc interview.

44. Dumping is defined as selling abroad at prices below fair market value in the home market.

45. Jeffrey A. Hart, "The Anti-Dumping Petition."

46. Ibid.

47. Norihiko Naono interview.

48. Norihiko Naono, "Japan FPD Market: Industry at Large," *Electronic News,* September 7, 1992, 10. This point was reiterated repeatedly in our research by most U.S. and all Japanese government officials, although some U.S. officials had the opposite impression.

49. In 1989, MITI encouraged FPD producers to form the Giant Technology Corporation with a funding level of about $28 million to pursue non-lithographic printing techniques to fabricate very large-format TFTs. The formula of cost sharing between business and government is subject to dispute. See Michael Borrus and Jeffrey A. Hart, "Display's the Thing"

and Norihiko Naono, "Japan FPD Market." According to IBM's Steven Depp, this program proved itself a major misdirection of corporate resources. Telephone discussion with Lenway and Murtha, November 14, 2000.

50. Martin Koughan, writer/producer/reporter, and David Ewing, director, "Losing the War with Japan," first broadcast on U.S. Public Television's *Frontline*, November 19, 1991.

51. Sheila Galatowitsch, "LCDs Run Away with Military Flat Panel Market," *Defense Electronics*, May 1993, 25.

52. "Pentagon Picks Partner for Flat Screens," *San Francisco Chronicle*, February 5, 1992, F3. The discussions reflected the companies' response, submitted January 18, 1993, to a request for proposals issued by DARPA nine months earlier.

53. "SEMI Forms FPD Division to Serve U.S. Display Consortium," Business Wire, July 20, 1993, accessed via Lexis-Nexis, summer 1994.

54. Ibid.

55. Ross Young, "Silicon Sumo."

56. IBM is a member of the consortium's user group, but not the producer group.

57. By 1998, projects budgeted at over $95 million had been funded in this way.

58. For a discussion of the critical role of practice in learning and knowledge creation, see John Seely Brown and Paul Duguid, *The Social Life of Information*, especially Chapters 4 and 5.

59. Interview, industry official with Murtha and Lenway, Silicon Valley, summer 1996.

60. Rex Tapp interview.

61. Scott Holmberg interview.

62. Photon Dynamics raised capital in Japan with the help of the Nomura and Daiwa Securities firms beginning in 1991. Bob Johnstone, "Research and Innovation: Spot the Mistake," *Far Eastern Economic Review*, July 16, 1992, 66. The Korean company LG also took a position in the company, according to Choon-Rae Lee (interview).

63. Various confidential interview materials including multiple government, industry, and company officials.

64. Jack Robertson, "A Closer Look: USDC Needs Foreign Aid in FPD Market," *Electronic Buyers' News*, January 24, 2000, 4.

65. "The Days of OIS," *The Clock*.

66. David Lieberman, "Xerox's dpiX Group Stumbles Six Months After Optical Imaging Systems Pulled Plug on Defense Displays—Military Reels as Latest LCD Maker Falters," *Electronic Engineering Times*, March 29, 1999, accessed via Lexis-Nexis, November 23, 2000.

67. "Planar Systems and Consortium Complete Purchase of dpiX; Third Quarter Earnings will not Meet Expectations," Business Wire, July 9, 1999, accessed via Lexis-Nexis, November 25, 2000.

68. "Planar Rebounds with Record Sales," *The Oregonian*, October 31, 2000, accessed via PanelX.com, November 23, 2000.

69. Jun H. Souk interview.

70. Jae-Choon Lim interview.

71. USDC board member interview by Lenway in Boston, Massachusetts, summer 1996.

72. Samsung and Corning established their first joint venture in 1973 to manufacture CRT glass. See www.samsungcorning.co.kr, accessed November 16, 1996.

73. Samsung website, www.sec.samsung.co.kr, accessed November 16, 1996.

74. Choon-Rae Lee interview.

75. "South Korea, Taiwan Firms Raid Japanese Staffs, Buy Technology," *Nikkei Weekly,* March 3, 1997, 20.

76. De facto industry standards permit shipping goods with as many as five defective pixels.

77. "LG to Supply $1 Billion Worth of TFT LCDs to Compaq," *Korea Economic Weekly,* December 12, 1996, accessed via Lexis-Nexis, June 1997.

78. "South Korea Close to Production of Large LCDs," *Dempa Shimbun,* August 31, 1996, 1.

79. "Display Technology to Buy Hyundai Displays," *Dempa Shimbun,* October 24, 1996, accessed via COMLINE, October 25, 1996; also interview materials.

80. Yoo Choon-sik, "Hyundai Elec in Up to $50 Bln Chip, LCD Deals," Reuters, November 15, 1999, accessed via AOL, November 17, 1999.

81. "Hyundai to Boost LCD Business," *The Korea Herald,* March 17, 2000, accessed at http://www.nikkeibp.asiabiztech.com, September 29, 2000.

82. "Apple Going Big-Time Flat," CNNfn, July 28, 1999, accessed at cnnfn.com/ 1999/07/28/technology/apple/, August 1999.

83. Yoo Choon-Sik, "Hyundai Elec," op.cit., note 80.

84. Alan Paterson, "What's Wrong with This Picture," *Electronic Business Asia,* April 2000, accessed at eb-asia.com, April 10, 2000.

85. Kam Law, "Flat Panel Display Manufacturing Challenges."

86. Choon-Rae Lee interview.

87. B.H. Seo, "Defiant LG Hints at Compromise," *Electronic Engineering Times,* January 4, 1999, accessed at www.edtn.com/news/0199/0199/010499bnews3.html, October 16, 1999.

88. See www.lgphilips-lcd.com.eng/company/lcd_history.html, accessed October 16, 1999.

## CHAPTER 9: KNOWLEDGE AND TRANSCENDENCE

1. Interview.

2. "Taiwan Boosts its Output of Liquid Crystal Displays," *Far Eastern Economic Review,* October 5, 2000, 37.

3. Chris Hall, "Bright Picture; Fuzzy Future," Electronic Business Asia, August 2000, accessed at eb-asia.comm, August 2000.

4. Mitchell Bernard and John Ravenhill adapted an analogy of "Flying Geese" made by Akamatsu Kaname in the 1930s to refer to this pattern of diffusion. See "Beyond Product Cycles."

5. See, for example, Jang-jin Hwang, "Memory Chip, TFT LCD Producers Suffer from Weak Sales, Prices in 2000," *The Korea Herald,* December 28, 2000, accessed at http://www.nikkeibp.asiabiztech.com, September 29, 2000.

6. Jang-Jeng Liang and Kuang-Lang Chen interview. See also Greg Linden, Jeffrey A. Hart, Stefanie Ann Lenway, and Thomas P. Murtha, "Flying Geese as Moving Targets."

7. Hsing C. Tuan, "Taiwan's Heavy Investments."

8. Faith Hung, "Asahi Joins Taiwan's TFT LCD Hub: Japanese Company to Construct $281M Glass Substrate Facility," *Electronic Buyers' News,* November 20, 2000, 10.

9. In December 2000, for example, *Wired* magazine's "Best" Department (382) reported on test evaluations of three PDP screens, a 50-inch (diagonal measurement) model by Pioneer for $19,995, a 42-inch Sony for $15,999, and a 42-inch Panasonic for $9,995.

10. Senior managers from several companies have declared that PDP will break through the price barrier in 2002. See, for example, Yoshiko Hara, "Joint-Venture Fab Aims to Usher in Plasma-TV Era," *Electronic Engineering Times,* October 2, 2000, accessed via Lexis-Nexis, January 2, 2001.

11. Colin Tan, "Hub for Advanced LCD Plants Ahead," *The Straits Times,* September 20, 2000, 3.

12. Yoonhee Park, "Korea May Halt Display Aid," *Electronic Engineering Times,* May 29, 2000, accessed via Lexis-Nexis, January 2, 2001; "Seoul to Give Tariff-Free Status to LCD Equipment Imports," *The Korea Herald,* December 9, 2000, accessed at http://www.nikkeibp. asiabiztech.com, September 29, 2000.

13. "Corning to Invest US$320 million in Taiwan's Glass Substrate Production," *Commercial Times,* Taiwan, September 19, 2000, accessed at
http://www.nikkeibp.asiabiztech.com, September 29, 2000.

14. Email communication, Robert L. Yard to Lenway and Murtha, December 1, 2000.

15. Email communications, Tetsuo Iwasaki to Lenway and Murtha, December 24 and 28, 2000.

16. "Toshiba to Withdraw from LCD Joint Venture with IBM," July 4, 2001, and "IBM Japan, Taiwan's Chi Mei to Set Up TFT LCD Company, July 6, 2001, AsiaBizTech, accessed at http://bizns. nikkeibp.co.jp/cgi-bin/asia/show/nsh_titlelist, July 23, 2001.

17. Suzanne Kapner, "LG of Korea and Philips Set Screen-Making Venture," *New York Times,* November 28, 2000, Section W, 1.

18. John Markoff, "Rivals Cooperate on Chip Equipment," *New York Times,* December 12, 2000, accessed at nytimes.com, December 12, 2000.

19. SEMATECH, Annual Report, Austin, Texas, 1999, accessed at sematech.org, January 2, 2001.

20. See Frank Rose, "Teleprompter: Europe Leads the Way in Advancing Wireless," *Wired,* December 2000, 331–333, for some exemplary statistics. See also Suh-kyung Yoon, "The Musty Smell of Success," *Far Eastern Economic Review,* January 25, 2001, 32–35.

21. Alicia Neumann, "Into Scandinavia," *Red Herring,* April 2000, 228–230; William Shaw, "In Helsinki Virtual Village," *Wired,* March 2001, 156–163.

22. Frank Rose, "Vivendi's High Wireless Act," *Wired,* December 2000, 318–330. Market forecasts quoted in the article originated with Jupiter Research.

23. Tom Stein, "Cutting the Cord," *Red Herring,* April 2000, 224–227.

24. Jean-Luc Grand-Clement interview. See also Yves L. Doz, Peter Smith Ring, Stefanie Lenway, and Thomas P. Murtha, "PixTech, Inc."

25. Tom Stein, "Cutting the Cord," op.cit., note 23.

26. The Swiss, Singaporean, Canadian, and Australian governments have already set up such missions, with similar ideas under consideration in the governments of Japan, the United Kingdom, Ireland, and Finland. See Lynnley Browning, "Coming to America to Learn a

Secret: Boldness." *New York Times,* December 10, 2000, accessed via NYT web archives on December 29, 2000.

27. Jack Robertson, "Flat Panel Makers Urged to Adopt Glass Standard," *Electronic Buyers' News,* December 15, 2000, accessed via Lexis-Nexis, February 25, 2001.

28. Barbara Jorgensen, "New, Small-Screen Technologies Display Some Promise," *Electronics Business,* October 2000, accessed via Lexis-Nexis, January 2, 2001.

29. Steven Depp, telephone conversation with Lenway and Murtha, February 1, 2001.

30. "Prime View to Move into Large-Scale TFT LCD Production," *Commercial Times,* Taiwan, September 18, 2000, accessed at http://www.nikkeibp.asiabiztech.com, September 29, 2000.

# REFERENCES

Abernathy, W.J. and J.M. Utterback. "Patterns of Industrial Innovation." *Technology Review* 80 (June/July 1978): 41–47.

Alt, Paul and Peter Pleshko. "Scanning Limitations of Liquid Crystal Displays." *IEEE Trans. Electron Devices* 21 (1963): 146.

Anderson, Sherwood. *Winesburg, Ohio.* Norton Critical Edition. Edited by Charles E. Modlin and Ray Lewis White. New York: W.W. Norton, 1996.

Arrow, Kenneth J. *Essays on the Theory of Risk Bearing.* Chicago: University of Chicago Press, 1971.

Arthur, W. Brian. *Increasing Returns and Path Dependence in the Economy.* Ann Arbor: University of Michigan Press, 1994.

———. "Industry Location Patterns and the Importance of History." In *Increasing Returns and Path Dependence in the Economy,* 49–68. Ann Arbor: University of Michigan Press, 1994.

———. "Increasing Returns and the New World of Business." *Harvard Business Review.* (July–August 1996): 100–109.

Arthur, W. Brian and David A. Lane. "Information Contagion." In *Increasing Returns and Path Dependence in the Economy,* 69–98. Ann Arbor: University of Michigan Press, 1994.

Bartlett, Christopher A. and Sumantra Ghoshal. *Managing Across Borders: The Transnational Solution.* Boston: Harvard Business School Press, 1989.

Basalla, George. *The Evolution of Technology.* New York: Cambridge University Press, 1988.

Bernard, Mitchell and John Ravenhill. "Beyond Product Cycles and Flying Geese: Regionalization, Hierarchy, and the Industrialization of East Asia." *World Politics* 47 (January 1995): 171–209.

Bitzer, Donald. "Inventing the AC Plasma Panel." *Information Display.* (February 1999): 22–27.

Borrus, Michael and Jeffrey A. Hart. "Display's the Thing: The Real Stakes in the Conflict over High Resolution Displays." *Journal of Policy Analysis and Management* 13 (Winter 1994): 21–54.

Brinkley, Joel. *Defining Vision: The Battle for the Future of Television.* New York: Harcourt Brace, 1997.

Brody, T. Peter. "The Birth and Early Childhood of Active Matrix: A Personal Memoir." *Journal of the Society for Information Display.* (April 1996): 113–127.

Brown, John Seely and Paul Duguid. *The Social Life of Information.* Boston: Harvard Business School Press, 2000.

Brown, Shona L. and Kathleen M. Eisenhardt. *Competing on the Edge: Strategy as Structured Chaos.* Boston: Harvard Business School Press, 1998.

Castellano, Joseph A. "Trends in the Global Flat Panel Display Industry." Presentation at the United States Display Consortium (USDC) Investors' Conference, Display Works, San Jose, California, February 6–8, 1996.

Caves, Richard. *The Multinational Corporation and Economic Analysis.* 2nd edition. Cambridge: Cambridge University Press, 1997.

Chandler, Alfred D. *Strategy and Structure: Chapters in the History of the American Industrial Enterprise.* Cambridge, Massachusetts: MIT Press, 1962.

_____. *Scale and Scope: The Dynamics of Industrial Capitalism.* Cambridge, Massachusetts: The Belknap Press of Harvard University Press, 1990.

Curry, James and Martin Kenney. "Beating the Clock: Corporate Responses to Rapid Change in the PC Industry." *California Management Review* 42 (Fall 1999): 8–36.

David, Paul. "Clio and the Economics of QWERTY." *American Economic Review Proceedings* 75 (1985): 332–337.

Depp, Steven W. "Presentation." *Flat Panel Display Strategic Forum Proceedings: Creating a U.S. Industry.* Ann Arbor: Center for Display Technology and Manufacturing, College of Engineering, The University of Michigan, November 15–16, 1994, 12.

Depp, Steven W. and Webster E. Howard. "Flat-Panel Displays: Recent Advances in Microelectronics and Liquid Crystals Make Possible Video Screens that can be Hung on a Wall or Worn on a Wrist." *Scientific American* (March 1993): 90–97.

Diamond, Jared. *Guns, Germs and Steel: The Fates of Human Societies.* New York: Norton, 1997.

DisplaySearch. *Equipment and Materials Analysis and Forecast.* Austin, Texas: DisplaySearch, 1999.

_____. "Flat Panel Fab Database." *Flat Panel Display Fabs on Disk.* Austin, Texas: DisplaySearch, June 1997.

Dosi, Giovanni. "Technological Paradigms and Technological Trajectories: A Suggested Interpretation of the Determinants and Directions of Technological Change." *Research Policy* 11 (3): 147–163.

Doz, Yves L., Peter Smith Ring, Stefanie Ann Lenway, and Thomas P. Murtha. "PixTech, Inc." Case 398-140-1. Fontainebleau, France: INSEAD, 1998.

Doz, Yves L., Jose Santos, and Peter Williamson. *From Global to Metanational: How Companies Win in the Knowledge Economy.* Boston: Harvard Business School Press, forthcoming.

ED Research of Japan. *Full Details of Japanese Liquid Crystal Manufacturing Lines.* Translators, InterLingua, Inc., Redondo Beach, California: Interlingua, 1994.

Ferrero, Lisa. "AMLCD Substrate Market." In *Proceedings of the DisplaySearch Economics of the Display Industry Conference,* Session VI. Austin, Texas: DisplaySearch, March 10–11 1999.

Florida, Richard and David Browdy. "The Invention that Got Away." *Technology Review* 93 (September–October 1991): 42–55.

Hager Grob, Daniela. "The Flat Panel Display Industry in Japan." Working paper. Tokyo: Sophia University. April 20, 1994 version, laserscript.

Ham, Rose Marie and David C. Mowery. "Enduring Dilemmas in U.S. Technology Policy." *California Management Review* 37 (Summer 1995): 89–107.

Hamel, Gary and Yves L. Doz. *Alliance Advantage: The Art of Creating Value Through Partnering.* Boston: Harvard Business School Press, 1998.

Hamel, Gary and C.K. Prahalad. *Competing for the Future: Breakthrough Strategies for Seizing Control of Your Industry and Creating the Markets of Tomorrow.* Boston: Harvard Business School Press, 1994.

Hart, Jeffrey A. "The Anti-Dumping Petition of the Advanced Display Manufacturers of America: Origins and Consequences." *The World Economy* 16 (January 1993): 85–109.

Harrison, Linden. "Liquid Crystal Displays." *Electronics Australia* (February 1973): 22–24. Reprinted from *Electron.*

Howard, Webster E. Foreword to *Color TFT Liquid Crystal Displays,* edited by Teruhiko Yamazaki, Hideaki Kawakami, and Hiroo Hori. Supervising translator, Shunsuke Obinata for O'Mara and Associates. Mountain View, California: Semiconductor Equipment and Materials International (SEMI), 1996.

Itoh, Joji and Seiichiro Yokoyama. "Structure of Color TFT Liquid Crystal Displays." Chapter 3 in *Color TFT Liquid Crystal Displays,* edited by Teruhiko Yamazaki, Hideaki Kawakami, and Hiroo Hori, 149–279. Principal translator, Shunsuke Obinata for O'Mara and Associates. Mountain View, California: Semiconductor Equipment and Materials International (SEMI), 1996.

Johnstone, Bob. *We Were Burning: Japanese Entrepreneurs and the Forging of the Electronic Age.* Boulder, Colorado: Basic Books (Westview Press), 1999.

Kidder, Tracy. *The Soul of a New Machine.* New York: The Modern Library, 1997.

Knox, Rick. "User Views of the FPD Industry." Presentation at "Three Views of the FPD Industry" panel, United States Display Consortium (USDC) Business Conference, Display Works, San Jose, California, February 5, 1996.

Kobrin, Stephen J. *Managing Political Risk Assessment: Strategic Response to Environmental Change.* Berkeley: University of California Press, 1982.

Kogut, Bruce. "Designing Global Strategies: Comparative and Competitive Value-Added Chains." *Sloan Management Review* (Summer 1985): 15–28.

———. *Country Competitiveness: Technology and the Organizing of Work.* New York: Oxford University Press, 1993.

Kuemmerle, Walter. "Building Effective R&D Capabilities Abroad." *Harvard Business Review* (March–April 1997): 61–70.

Kuhn, Thomas. *The Structure of Scientific Revolutions.* 2nd Edition. Chicago: University of Chicago Press, 1970.

Law, Kam. "Flat Panel Display Manufacturing Challenges." In *DisplaySearch U.S. FPD 2000 Conference Proceedings,* CD-ROM, Session 4. Austin, Texas: DisplaySearch, March 21–22, 2000.

Lee, Dong Ho. *Foreign Market Entry: An Evolutionary Approach.* University of Michigan Business School Ph.D. dissertation. Ann Arbor: University Microfilms (UMI), 1991.

Lenway, Stefanie Ann and Thomas P. Murtha. "The State as Strategist in International Business Research." *Journal of International Business Studies* (Third Quarter 1994): 513–536.

Linden, Greg, Jeffrey Hart, Stefanie Ann Lenway, and Thomas P. Murtha. "Flying Geese as Moving Targets: Are Korea and Taiwan Catching Up with Japan in Advanced Displays?" *Industry and Innovation* 5 (June 1998): 11–34.

Maclean, Norman. *Young Men and Fire.* Chicago: University of Chicago Press, 1992.

Masters, Edgar Lee. *Spoon River Anthology.* New York: New American Library, 1992.

Morgan, James C. and J. Jeffrey Morgan. *Cracking the Japanese Market: Strategies for Success in the New Global Economy.* New York: Free Press, 1991.

Murtha, Thomas P. "Surviving Industrial Targeting: State Credibility and Public Policy Contingencies in Multinational Subcontracting." *Journal of Law, Economics and Organization* 7 (Spring 1991): 117–143.

———. "Credible Enticements: Can Host Governments Tailor Multinational Firms' Organizations to Suit National Objectives?" *Journal of Economic Behavior and Organization* 20 (February 1993): 171–186.

Murtha, Thomas P. and Stefanie Ann Lenway. "Country Capabilities and the Strategic State: How National Political Institutions Affect Multinational Corporations' Strategies." *Strategic Management Journal* 15 (Special Issue, Summer 1994): 113–129.

Nelson, Richard R. *National Innovation Systems: A Comparative Analysis.* New York: Oxford University Press, 1993.

Nikkei Business Press. *Nikkei Microdevices Flat Panel Display Yearbook.* Translators, Inter-Lingua.com, Inc. Tokyo: Nikkei Business Press, various years.

Noda, Tomo. "Sharp Corporation: Corporate Strategy." Case N9-793-064. Version of June 11, 1993. Boston: Harvard Business School, 1993.

Nonaka, Ikujiro and Hirotaka Takeuchi. *The Knowledge Creating Company: How Japanese Companies Create the Dynamics of Innovation.* New York: Oxford University Press, 1995.

Numagami, Tsuyoshi. "Strategic Deviation from the Trend of Technological Evolution: A Case Analysis of the Liquid Crystal Display Industry." Tokyo: Hitotsubashi University Graduate School of Commerce, 1987, mimeo.

Numakami, Kan, Ikujiro Nonaka, and Takebi Ohtsubo. "Sharp Technology Management." Case SMIP-91-16 (CN). Tokyo: Nomura School of Advanced Management, 1991.

O'Mara, William C. "Active Matrix Liquid Crystal Displays Part I: Manufacturing Process." *Solid State Technology* (December 1991): 65–70.

———. "Update on Flat Panel Displays." *Solid State Technology* (November 1993): 35–41.

———. "How to Make Money in the Display Business." Presentation at Display Forecast Roundtable, United States Display Consortium (USDC) Business Conference, Display Works, San Jose, California , February 5, 1996.

Polanyi, Michael. *Personal Knowledge: Towards a Post-Critical Philosophy.* Chicago: University of Chicago Press, 1962.

———. *The Tacit Dimension.* Gloucester, Massachusetts: Peter Smith, 1983.

Porter, Michael E. *Competitive Advantage: Creating and Sustaining Superior Performance.* New York: Free Press, 1985.

_____. "Competition in Global Industries: A Conceptual Framework." In *Competition in Global Industries,* edited by Michael E. Porter, 15-60. Boston: Harvard Business School Press, 1986.

_____. *The Competitive Advantage of Nations.* New York: The Free Press, 1990.

_____. "Clusters and the Economics of Competition." *Harvard Business Review* (November–December 1998): 77–90.

Prahalad, C.K. and Yves L. Doz. *Multinational Mission: Balancing Local Demands and Global Vision.* New York: Free Press, 1987.

Rosenbloom, Richard S. and Michael A. Cusumano. "Technological Pioneering: The Birth of the VCR Industry." *California Management Review* 29 (Summer 1987): 51–76.

Saccocio, Damian. "Strategy and New Business Development: The Case of the Missing U.S. Display Industry." Ph.D. diss., Rensselaer Polytechnic Institute, Troy, New York, 1994.

Scheffer, Terry J. "The 'Iron Law' 25 Years Later." *Information Display* (October 1998): 12–16.

Smith, Robert B. "Electronics Developments for Field Emission Displays." *Information Display* (February 1998): 12–15.

Sobel, Alan. "Television's Bright New Technology." *Scientific American* (May 1998): 70–77.

Stowsky, Jay. "America's Technical Fix: The Pentagon's Dual Use Strategy, TRP and the Political Economy of U.S. Technology Policy in the Clinton Era." Discussion Paper, Berkeley Roundtable on the International Economy (BRIE). Berkeley: BRIE, June 1996.

Teece, David J. "Profiting from Technological Innovation." In *The Competitive Challenge: Strategies for Industrial Innovation and Renewal,* edited by David J. Teece, 185–219. Cambridge, Massachusetts: Ballinger, 1987.

Tuan, Hsing C. "Taiwan's Heavy Investments in TFT-LCD Manufacturing." In *Proceedings of the DisplaySearch Economics of the Display Industry Conference,* Session II. Austin, Texas, DisplaySearch, March 10–11, 1999.

United States Congress. Office of Technology Assessment. *The Big Picture: HDTV and High-Resolution Systems.* Washington, D.C.: USGPO, June 1990.

United States Department of Defense. *Building U.S. Capabilities in Flat Panel Displays: Final Report of the Flat Panel Display Task Force.* Washington, D.C.: DoD, October 1994.

Vernon, Raymond. *Sovereignty at Bay: The Multinational Spread of U.S. Enterprises.* New York: Basic Books, 1971.

_____. *In the Hurricane's Eye: The Troubled Prospects of Multinational Enterprises.* Cambridge, Massachusetts: Harvard University Press, 1998.

Wakabayashi, Hideki. "Liquid Crystal Displays: The Next Five Years." *NRI Quarterly* (Summer 1995): 62–85.

Waldrop, M. Mitchell. *Complexity: The Emerging Science at the Edge of Order and Chaos.* New York: Simon and Schuster, 1992.

Weick, Karl. *Sensemaking in Organizations.* Thousand Oaks, California: Sage, 1995.

_____. "Prepare Your Organization to Fight Fires." *Harvard Business Review* (May–June 1996): 143–148.

West, Jonathan and H. Kent Bowen. "Display Technologies, Inc." Case 9-699-006. Version of July 28, 1998. Boston: Harvard Business School, 1998.

Yamazaki, Teruhiko, Hideaki Kawakami, and Hiroo Hori, ed. *Color TFT Liquid Crystal Displays.* Principal translator Shunsuke Obinata for O'Mara Associates. Mountain View, California: Semiconductor Equipment and Materials International (SEMI), 1996.

Yoshino, Michael Y. and U. Srinivasa Rangan. *Strategic Alliances: An Entrepreneurial Approach to Globalization.* Boston: Harvard Business School Press, 1995.

Young, Ross. *Silicon Sumo: U.S.-Japan Competition and Industrial Policy in the Semiconductor Equipment Industry.* Austin: IC² Institute, University of Texas, 1994.

_____. "Current FPD Market Status." In *Proceedings of the DisplaySearch Economics of the Display Industry Conference,* Session I. Austin, Texas: DisplaySearch, March 10–11, 1999.

_____. "FPD Equipment and Materials Market Trends and Forecast." In *Proceedings of the DisplaySearch Economics of the Display Industry Conference,* Session VI. Austin, Texas: DisplaySearch, March 10–11, 1999.

_____. "FPD Market Status and Outlook." *DisplaySearch U.S. FPD 2000 Conference Proceedings,* CD-ROM, Session I. Austin, Texas: DisplaySearch, March 21–2, 2000.

# INDEX